Forty-First Annual International Pittsburgh Energy and Carbon Management Conference (PECMC 2024)

Technology Innovation and Transition to Cleaner Energy

AA005458

Online
7-9 October 2024

ISBN: 979-8-3313-2505-3

Printed from e-media with permission by:

Curran Associates, Inc.
57 Morehouse Lane
Red Hook, NY 12571

Some format issues inherent in the e-media version may also appear in this print version.

Printed with permission by Curran Associates, Inc. (2025)

For permission requests, please contact International Pittsburgh Coal Conference
at the address below.

International Pittsburgh Coal Conference
University of Pittsburgh
Swanson School of Engineering
3700 O'Hara St.
940 Benedum Hall
Pittsburgh, Pennsylvania 15261
USA

Phone: (412) 624-7440
Fax: (412) 624-1480

ipcc@pitt.edu

Additional copies of this publication are available from:

Curran Associates, Inc.
57 Morehouse Lane
Red Hook, NY 12571 USA
Phone: 845-758-0400
Fax: 845-758-2633
Email: curran@proceedings.com
Web: www.proceedings.com

Plenary Sessions

MONDAY, October 7, 2024

8:00-10:00 - PLENARY SESSION I

Moderator: Dr. Massood Ramezan

8:00-9:00 - **The Honorable Brad Crabtree**

Assistant Secretary for Fossil Energy and Carbon Management

"Carbon Management: Progress to Date and Challenges Going Forward"...N/A

9:00 - 10:00 - **Dr. Omar Basha**

Managing Director, ONB Engineering Research and Technical Services LLC

"Highlighting Critical Barriers in the CCUS Value Chain to Develop technically and Economically Feasible Solutions "...1

TUESDAY, October 8, 2024

8:00-10:00 - PLENARY SESSION II

Moderator: Mr. Thomas Sarkus

8:00 – 9:00 - **Dr. Ahmed Aboudheir**

Chief Executive Officer of Aboudheir Consulting Ltd

"Design and Deployment of Large Scale CCUS Projects that Work Towards Net Zero by 2050"...N/A

9:00 - 10:00 - **Dr. Marianne Walck**

Director of the National Energy Technology Laboratory (NETL) US-DOE

"Driving Clean Energy: NETL's Role, Integrated Systems, and the Path to Carbon Management"...N/A

WEDNESDAY, October 9, 2024

8:00-10:00 - PLENARY SESSION III

Moderator: Mr. Richard Winschel

8:00 – 9:00 - **The Honorable Charles McConnell**

Assistant Secretary of Energy (2011-13) US-DOE Energy Officer UH Energy

Executive Director, Carbon Management and Energy Sustainability Chancellor/President's Division

University of Houston

"The Energy Transition Requires CCUS for Transformation"...19

9:00 - 10:00 - **Dr. Evan Granite**

Program Manager, Carbon Ore

Senior Technical Advisor, U.S. DOE - Fossil Energy & Carbon Management

"Domestic Wastes and Byproducts: A Resource for Critical Material Supply Chains"...N/A

Times listed are Eastern Daylight Savings Time USA (Local Pittsburgh, PA, USA)

ORAL SESSIONS
Monday, October 7
10:00 —15:00

SESSION 1

CARBON MANAGEMENT — 1

Bingyun Li

10:00 Photothermal reduction of CO2 to formic acid over Pd/P25 composite catalyst, Fangyuan Zhang, Jieying Jing, Wen-Ying Li, Taiyuan University of Technology, Taiyuan, CHINA...N/A

10:30 Characteristic of hydrogen production from various of feed gases in sorption enhanced water gas shift reactions, Jinpeng Zhang, Yanxin Jia, Jieying Jing, Wenying Li, Taiyuan University of Technology, Taiyuan, CHINA...N/A

11:00 Adsorption and desorption kinetics of dry sorbents for CO_2 capture, Gregory N. Okolo, Juan Williams, Raymond C. Everson, John R. Bunt, Paola Ammendola, Federica Raganati, North-West University, SOUTH AFRICA...N/A

11:30 Lessons Learned from Canadian CCUS Projects, Laura Westover, Puttipong Tantikhajorngosol, The International CCS Knowledge Centre, Regina, SK, CANADA...25

SESSION 2

VALUE-ADDED PRODUCTS FROM COAL —1

Richard Winschel

10:00 Comparing ashes of some thermal power plants and metallurgical factory in Vietnam and their utilization, Nguyen Viet Quang Hung, Centre of Severe Materials and Environment, Hanoi, Vietnam; Masakatsu Nomura, Osaka University, Osaka, Japan; Chu Thi Thanh, TMD Group, Vinh City, Nghe An, Vietnam; Tadanori Hashimoto, Atsushi Ishihara, Mie University, Tsu - City, Mie, Japan...N/A

10:30 Application of composite external donor technology in coal-based polypropyle production, Peiqian Yu, Yufei Bian, Liuhia Feng, Yue Yu, Rongjuan Cong Lin feng, Chen, National institute of Clean-and-Low-Carbon, Beijing, CHINA...33

11:00 Molecular design and performance modulation of anthracene-based solar thermal fuels, Chonghua Li, Xingtang Xu, Jieying Jing, Wen-Ying Li, Taiyuan University of Technology, Taiyuan, CHINA

11:30 Efficient separation of monoaromatics from direct coal liquefaction, Xiao-Long Shi, Yun-Cai Song, Jie Feng, Wen-Ying Li, Taiyuan University of Technology, Taiyuan, CHINA...42

SESSION 3

CARBON MANAGEMENT — 2

Omar Basha

13:00 Converting CO_2 into Nano-Minerals via Amino Acids, Bingyun Li, Qingyang Li, Department of Orthopaedics, West Virginia University, Morgantown, WV, USA...N/A

13:30 Carbon Management Driven by Wafer Scale Engines, Vyacheslav Romanov, National Energy Technology Laboratory (NETL), Pittsburgh, PA, USA...51

14:00 Direct Air Capture Deployment Assessment in Deep Mitigation Pathways Toward 2075 in the U.S.: MARKAL-NETL Modeling Results, Christopher Nichols, Nadejda Victor, National Energy Technology Laboratory, USA...N/A

SESSION 4

VALUE-ADDED PRODUCTS FROM COAL - 2

Richard Bajura

13:00 NOVEL PRB COAL BASED ASPHALT, L.C. Muller, Cheng-Yen Pan, T.I. Pfeiffer, P. Behrens, J.J. Adams, Y. Kumbargeri, J.-P. Planche, J. Martin, L. Hazard, Md I. Haque, University of Wyoming - School for Energy Resources, Western Research Institute, Laramie, WY, USA...N/A

13:30 A Novel Approach to Manufacturing Lignite Composites with High-Filler Content, Md. Reza-E-Rabby, Keerti Kappagantula, Jose Ramos, Raveen John, Aditya Nittala, Yelin Li, Nathan Canfield, Adam Denny, Ethan Nickerson, Satish Nune, David J Heldebrant, Pacific Northwest National Laboratory, Richland, WA, USA...N/A

14:00 ROCK WOOL SPUN FROM VITRIFIED COAL FLY ASH/SLAG IS VERY PROFITABLE, Howard McClintic, Concurrent Technologies Corporation, Washington, DC, USA...58

14:30 Critical Elements for the Construction of a sustainable Closure* for CCR Impoundment, Mark Keck, Cristine Zubrow, Al Briggs, Key Environmental, Inc., Carnegie, PA. USA...N/A

ORAL SESSIONS
Tuesday, October 8
10:00 —15:00

SESSION 5

COAL SCIENCE/CLEAN COAL AND GAS TO FUELS

Isaac Gamwo

10:00 Evaluation of temporal variations in the chemistry of exposed coal heaps, Jyoti Sharma, Ashok K Singh, Arnold L Usham, Priya Kumari, CSIR-Central Institute of Mining and Fuel Research, Dhanbad (DC), Jharkhand, INDIA; Academy of Scientific and Innovative Research (AcSIR), Ghaziabad, INDIA...N/A

10:30 Preparation And Catalytic Cracking Reactivity of Zsm-5 Zeolite from Vietnamese Quang Ninh Coal Ash, Atsushi Ishihara, Hung Viet Nguyen, Koki Kunieda, Shinya Matsuura, Tadanori Hashimoto, Mie University, Tsu City, Mie Pref., JAPAN...N/A

11:00 Enhanced Phenanthrene Hydrogenation Saturation Performance over Nickel-Based Catalyst via Adjusting the Interaction Between Ni and Support, Tao Tian, Zemin Zhao, Jieying Jing, Wenying Li, State Key Laboratory of Clean and Efficient Coal Utilization, Taiyuan University of Technology, CHINA...N/A

11:30 Blue Ammonia and Urea Production via Coal-Gasification, Ravi Ravikumar, Sunil Kumar Singhal, Ben Vivier, Fluor Enterprises, Aliso Viejo, CA, USA...62

SESSION 6

GASIFICATION TECHNOLOGIES

Massood Ramezan

10:00 The formation and evolution of pores during the pyrolysis of low-rank coal and its influence on products mass transfer, Yi Cui, Yun-Cai Song, Jie Feng, Wen-Ying Li, State Key Laboratory of Clean and Efficient Coal Utilization, Taiyuan University of Technology, Taiyuan, CHINA...N/A

10:30 Building an Energy System Based on Thar Coal, Farid A. Malik, Abid Aziz, FCC University, Lahore, Pakistan; Pak Motors (Pvt) Ltd, PAKISTAN...70

11:00 U.S. Department of Energy/ National Energy Technology Laboratory Gasification Systems Program Overview, Jonathan W. Lekse, U.S. DOE, National Energy Technology Laboratory, Pittsburgh, PA, USA...N/A

11:30 Machine Learning Enhanced LIBS to Measure and Process Biofuels and Waste Coal for Gasifier Improved Operation, Ozan Karadut, Zheng Yao, Carlos Romero, Robert De Saro, Joe Craparo, Larry Mulligan, Johan Van Dyk, Asmara Soomro, Alexander Koutsostamatis, Lehigh University, Energy Research Center; GTI Energy, Energy Research Company, USA...75

SESSION 7

CARBON MANAGEMENT — 3

Omar Basha

13:00 Advancements in Methane-to-Methanol Conversion Technologies: A Review, Omar M. Basha, Rui Wang, Isaac Gamwo, Badie I. Morsi, ONB Engineering Research and Technical Services LLC, Delaware, USA; U.S. DOE NETL, Pittsburgh, PA, USA; University of Pittsburgh, Pittsburgh, PA, USA...90

13:30 TEA of the SMR-CCS Process for Blue Hydrogen Production Using Thirty-Six Physical Absorbents, Rui Wang, Husain E. Ashkanani, Nicholas S. Siefert, Robert L. Thompson, Jeffrey Culp, Kathryn Smith, Janice A. Steckel, Wei Shi, Isaac K. Gamwo, David Hopkinson, Victor Kasuma, Badie I. Morsi, U.S. DOE National Energy Technology Laboratory, Pittsburgh, PA, USA; Kuwait University, Kuwait; University of Pittsburgh, Pittsburgh, PA, USA....98

14:00 Performance of Anthracite Coal-Derived Activated Carbon in Adsorbing CO_2 across Different Gas Concentrations, Zeyuan Tian, Gabriel Bonsall, Zheng Yao, Carlos Romero, Tom Lowe, Greg Driscoll, Boyd Kreglow, Harold Schobert, Lehigh University - Energy Research Center, Blaschak Anthracite Corporation, USA...122

14:30 Impact of a bluff-body on pressurized oxy-combustion, Lei Li, V'yacheslav Akkerman, WVU, Morgantown, West Virginia, USA; Richard L. Axelbaum, Duarte Magalhaes, Mao Cheng, Washington University in Saint Louis, MO, USA...156

SESSION 8

CLEAN COAL DEMONSTRATION AND COMMERCIAL PROJECTS

Thomas Sarkus

13:00 A Random Walk Through...NETL's Commercial Success Stories, Thomas A. Sarkus, National Energy Technology Laboratory, USA...N/A

13:30 Planned Unit Will Produce REE Alloys for Permanent Magnets, Howard McClintic, Concurrent Technologies Corporation, Washington, DC, USA...164

14:00 Status of DOE Carbon Management Demonstration and Large Pilot Plant Projects, Thomas Sarkus, National Energy Technology Laboratory, U.S. Department of Energy, USA...N/A

ORAL SESSIONS
Wednesday, October 9
10:00 —15:00

SESSION 9

CLEAN HYDROGEN

Thomas Sarkus

10:00 Design of Non-Fluorinated Proton Exchange Membranes from Poly(Terphenyl Isatin) with Fluorene-Linked Sulfonate Groups and Microblock Structures, Chaoyi Ba, Shu Xu, Meltem Urgun-Demirtas, Argonne National Laboratory, Lemont, IL, USA...N/A

10:30 Thermal-catalytic Decomposition of Natural Gas for Clean Hydrogen and premium Carbon Using Electric Fields, Randy Vander Wal, James Heim, II, Mpila Nkiawete, Penn State University, University Park, USA...N/A

11:00 Performance Testing of a Moving -Bed Gasifier Using Coal Biomass and Waste Plastic Blends with Washed and Unwashed Legacy Coals, Horst Hack, George Booras, Rolf E. Maurer, David P. Thimsen, Elijah Thimsen, Alberto Pettinau, Simone Meloni, Gabriele Calli, Hang Zhou, Yupeng Xu, Mehrdad Shahnam, Electric Power Research Institute (EPRI), USA, Hamilton Maurer International, Inc. Hudson, IL, USA, Sotacarbo Sotacarbo Società Tecnologie Avanzate Low Carbon S.p.A., Carbonia, ITALY, National Energy Technology Laboratory (NETL), Morgantown, WV, USA...N/A

11:30 Halo: Hydrogen-Recovery Using an Ai-Arc-Plasma Learning Operational System For Produced Water, Glen Nakafuji, Oceanit, Honolulu, HI, USA...N/A

SESSION 10

COAL BED METHANE AND SHALE GAS/ SUSTAINABILITY & ENVIRONMENT

Richard Bajura

10:00 Analyzing Thanh Hoa, Ha Tinh and Nghe An coal samples of Vietnam, Nguyen Viet Quang Hung, Centre of Severe Materials and Environment, Hanoi, Vietnam; Masakatsu Nomura, Osaka University, Osaka, Japan; Chu Thi Thanh, TMD Group, Vinh City, Nghe An, Vietnam; Tadanori Hashimoto, Atsushi Ishihara, Mie University, Tsu - City, Mie, Japan...N/A

10:30 Research Progress and Key Scientific Issues of Unconventional Natural Gas, Dameng Liu, China University of Geosciences, Beijing, CHINA...N/A

11:00 Understanding the Regional Variations in Pay Thickness of Unconventional Reservoirs in the Powder River Basin by Incorporating Learnings from a Site-Specific, Core-Calibrated, Geologic Model, Luke Fritz, Richard Ness, Ben Roth, Nick Jones, Advanced Resources International Advanced Resources International, TN; Enhanced Oil Recovery Institute, WY, USA...169

11:30 Road Traffic Density Analysis and Multi-Objective Optimization of Coal Hauling Trucks Selection by Considering Cost and CO2 Emissions towards Cleaner Mining Production: Case Study of PT ABC, Irfan Aji Ramadzan, Firly Rachmaditya Baskoro, Tri Karian, Fadhila Achmadi Rosyid, Arjo Prawoto Wibowo, Institut Teknologi Bandung, INDONESIA...182

SESSION 11

RARE EARTH ELEMENTS/ENERGY STORAGE

Richard Winschel

13:00 Photoelectrocatalytic Hydrogen Production from Seawater Using Element-Doped TiO_2 Nanorods, Chen-Hang HuangFu, Yun-Cai Song, Jie Feng, Wen-Ying Li, State Key Laboratory of Clean and Efficient Coal Utilization, Taiyuan University of Technology, Taiyuan, CHINA...N/A

13:30 Beyond Combustion – Coal in the 21st Century – Carbon Products from Coal, Evan J. Granite, U.S. DOE, USA...N/A

14:00 Critical Mineral Supplies and Market Sector Demands for LIBs and EVs, Randy Vander Wal, James Heim, Penn State University, University Park, PA, USA...N/A

SESSION 12

COMBUSTION TECHNOLOGIES

Evan Granite

13:00 Life Cycle Greenhouse Gas Emissions of Coal-Biomass Co-firing Power Plants with Carbon Capture and Storage, Wanying Wu, Haibo Zhao, University of Wyoming, Laramie, WY, USA...N/A

13:30 Thermogravimetric Analysis Combustion Characteristics of Coal Microlithotype Particle Types, Lesigen Moodley, Frikkie Conradie, Walter Schmitz, Hein Neomagus, Bonny Nyangwa, Olifant Sinthumule, Eskom Research, Testing and Development, Johannesburg; North-West University; University of Witwatersrand, SOUTH AFRICA...188

14:00 The effect of dry salt conditioning on the resistivity of South African fly ashes, Gregory N. Okolo, Barend Mienie, Burgert B. Hattingh, Hein W.J.P. Neomagus, North-West University, SOUTH AFRICA...N/A

14:30 Drop Tube Furnace Combustion Characteristics and Computation Fluid Dynamics Modelling Of Coal Char Microlithotype Particle Types, Lesigen Moodley, Frikkie Conradie, Walter Schmitz, Hein Neomagus, Bonny Nyangwa, Kganuwi Kekana, Eskom Research, Testing and Development, Johannesburg; North-West University; University of Witwatersrand, SOUTH AFRICA...205

Forty-First Annual International Pittsburgh Energy and Carbon Management Conference (PECMC) (Virtual)

October 07-09, 2024

Technical Program

"Technology Innovation and Transition to Cleaner Energy"

Participating Organizations

Ajou University, SOUTH KOREA
Commonwealth Scientific and Industrial Research Organisation (CSIRO), AUSTRALIA
CONSOL Energy Inc., USA
Freiberg University of Mining & Technology, GERMANY
Gas Technology Institute (GTI), USA
Hamilton Maurer International, USA
HTI, USA
Institute for Chemical Processing of Coal, POLAND
Istanbul Technical University, TURKEY
KeyLogic Systems, Inc., USA
Longbridge Energy Consulting, USA
MIE University, JAPAN
National Institute of Clean and Low-Carbon Energy (NICE), CHINA
Osaka University, JAPAN
Penn State University, USA
Sotacarbo S.p.A., ITALY
Southern University of Science and Technology, CHINA
The Ohio State University, USA
U.S. DOE/NETL, USA
U.S. Geological Survey, USA
University of Houston, USA
University of Kentucky, USA
University of New Orleans, USA
University of Pittsburgh, USA
University of Wyoming, USA
West Virginia University, USA

Hosted by: University of Pittsburgh

WELCOME!

On behalf of the Conference Advisory Board, Conference Committees, and the University of Pittsburgh, we welcome you to the Forty-first Annual International Pittsburgh Energy and Carbon Management Conference (PECMC-2024), which will be held October 7-9, 2024 online via Zoom. The Conference is hosted by the University of Pittsburgh, USA.

The theme of this year's conference "Technology Innovation and Transition to Cleaner Energy" covers wide a spectrum of important topics on energy and environmental issues and technologies, which are directly related to secure transition to clean energy. This year's Technical Program includes two parallel oral tracks on Monday (10/7/2024), two parallel oral tracks on Tuesday (10/8/2024), and two parallel oral tracks on Wednesday (10/9/2024). All tracks begin at 10:00 and end at 15:00 each day. Detailed information about the technical sessions are available in page 8 of the Technical Program.

This year's invited Plenary Speakers are: (1) The Honorable Brad Crabtree, Assistant Secretary for Fossil Energy and Carbon Management; (2) Dr. Marianne Walck, Director of the National Energy Technology Laboratory (NETL) US-DOE; (3) The Honorable Charles McConnell, Assistant Secretary of Energy (2011-13) US-DOE, Energy Officer UH Energy, Executive Director, Carbon Management and Energy Sustainability Chancellor/President's Division, University of Houston; (4) Dr. Ahmed Aboudheir, Chief Executive Officer of Aboudheir Consulting Ltd; (5) Dr. Evan Granite, Program Manager, Carbon Ore, Senior Technical Advisor, U.S. DOE - Fossil Energy & Carbon Management; and (6) Dr. Omar Basha, Managing Director, ONB Engineering Research and Technical Services LLC.

We express our sincere gratitude to the Plenary Speakers and Moderators for their support and involvement, to all the authors and co-authors of the technical papers and to all the members of the Technical Program Committee. Special thanks go to the topic coordinators, co-chairs and speakers for their invaluable contributions to the 2024 technical program.

As the chair of the Advisory Board of the Conference, I deeply appreciate your participation and interest in this year's Conference and we invite you to join us next year for the Forty-second Annual International Pittsburgh Energy and Carbon Management Conference (PECMC-2024), hopefully in-person, at the Westin Convention Center Hotel in Pittsburgh, PA, USA .

Sincerely,

Massood Ramezan, Chair

M. Ramezan

Senior Technical Advisor at Key Logic Systems, USA

The International Pittsburgh Energy and Carbon Management Conference

EXECUTIVE DIRECTOR:
Dr. Badie I. Morsi

CONFERENCE OFFICE:
University of Pittsburgh, Swanson School of Engineering
Pittsburgh Coal Conference
3700 O'Hara Street, 940 Benedum Hall, Pittsburgh, PA 15261 USA
Tel: +1-412-624-7440 Email: ipcc@pitt.edu www.pccpitt.org

CONFERENCE OVERVIEW

MONDAY, OCTOBER 7, 2024

PLENARY SESSION I	08:00 - 10:00
CONCURRENT TECH. SESSIONS	10:00 - 12:00
LUNCH	12:00 - 13:00
CONCURRENT TECH. SESSIONS	13:00 - 15:00

TUESDAY, OCTOBER 8, 2024

PLENARY SESSION II	08:00 - 10:00
CONCURRENT TECH. SESSIONS	10:00 - 12:00
LUNCH	12:00 - 13:00
CONCURRENT TECH. SESSIONS	13:00 - 15:00

WEDNESDAY, OCTOBER 9, 2024

PLENARY SESSION IIII	08:00 - 10:00
CONCURRENT TECH. SESSIONS	10:00 - 12:00
LUNCH	12:00 - 13:00
CONCURRENT TECH. SESSIONS	13:00 - 15:00

Times listed are Eastern Daylight Savings Time USA (Local Pittsburgh, PA, USA)

ZOOM INVITATIONS

	Track 1 (8:00-14:30)	Track 2 (10:00-15:00)
Monday 10/7/2024	Plenary Session I and Sessions 1 & 3 **Join Zoom Meeting** https://pitt.zoom.us/j/92309437229 **Meeting ID: 923 0943 7229** **Passcode: PECMC2024** **One tap mobile** **+12678310333,,92309437229# US (Philadelphia)** **8778535247,,92309437229# US Toll-free** **Find your local number: https://pitt.zoom.us/u/acik1aWEzE**	Sessions 2 & 4 **Join Zoom Meeting** https://pitt.zoom.us/j/96932055493 **Meeting ID: 969 3205 5493** **Passcode: PECMC2024** **One tap mobile** **+12678310333,,96932055493# US (Philadelphia)** **8778535247,,96932055493# US Toll-free** **Find your local number: https://pitt.zoom.us/u/aee50qZqBg**
	Track 1 (8:00-15:00)	**Track 2 (10:00-14:30)**
Tuesday 10/8/2024	Plenary Session II and Sessions 5 & 7 **Join Zoom Meeting** https://pitt.zoom.us/j/98781954319 **Meeting ID: 987 8195 4319** **Passcode: PECMC2024** **One tap mobile** **+12678310333,,98781954319# US (Philadelphia)** **8778535247,,98781954319# US Toll-free** **Find your local number: https://pitt.zoom.us/u/almUsy7qA**	Sessions 6 & 8 **Join Zoom Meeting** https://pitt.zoom.us/j/92512980106 **Meeting ID: 925 1298 0106** **Passcode: PECMC2024** **One tap mobile** **+12678310333,,92512980106# US (Philadelphia)** **8778535247,,92512980106# US Toll-free** **Find your local number: https://pitt.zoom.us/u/acDmBabPbf**
	Track 1 (8:00-14:30)	**Track 2 (10:00-15:00)**
Wednesday 10/9/2024	Plenary Session III and Sessions 9 & 11 **Join Zoom Meeting** https://pitt.zoom.us/j/91903279767 **Meeting ID: 919 0327 9767** **Passcode: PECMC2024** **One tap mobile** **+12678310333,,91903279767# US (Philadelphia)** **8778535247,,91903279767# US Toll-free** **Find your local number: https://pitt.zoom.us/u/apGSAPFb4**	Sessions 10 & 12 **Join Zoom Meeting** https://pitt.zoom.us/j/96038226849 **Meeting ID: 960 3822 6849** **Passcode: PECMC2024** **One tap mobile** **+12678310333,,96038226849# US (Philadelphia)** **8778535247,,96038226849# US Toll-free** **Find your local number: https://pitt.zoom.us/u/adQMIB6DLD**

Times listed are Eastern Daylight Savings Time USA (Local Pittsburgh, PA, USA)

Virtual Conference FAQ's

All participants are expected to conduct themselves in a professional and respectful manner during the conference and abide by all applicable University policies and procedures. While this conference is a public discussion, participants may not record any session.

How do I join a session?

Conference staff will email the Zoom meeting details to all paid conference attendees the day before the conference. It is recommended that you test Zoom before the conference.

How do I ask a question during a session?

To submit a question to the speaker, type it into the Zoom chat box. The session co-chair will read questions from the chat to the speaker.

I am giving an oral presentation, what do I need to do during my session?

Please sign in to the Zoom session 10-15 minutes before your scheduled presentation time. At the start of your scheduled presentation time the session co-chair will announce your presentation and conference staff will play your pre-recorded video presentation or you will present live for 25 min. Following the talk there will be 5 minutes of live question and answer. The session co-chair will read you any questions submitted in the chat

Oral Presentation Sessions

	Monday (10/7/2024)	
	Track 1	**Track 2**
8:00 - 9:00	**Plenary Session I**	
9:00 - 10:00		
10:00 - 10:30	Carbon Management - 1 (Session 1)	Value-Added Products from Coal - 1 (Session 2)
10:30 - 11:00		
11:00 - 11:30		
11:30 - 12:00		
12:00- 13:00	**Lunch**	
13:00 - 13:30	Carbon Management - 2 (Session 3)	Value-Added Products from Coal - 2 (Session 4)
13:30 - 14:00		
14:00 - 14:30		
14:30 - 15:00		

	Tuesday (10/8/2024)	
	Track 1	**Track 2**
8:00 - 9:00	**Plenary Session II**	
9:00 - 10:00		
10:00 - 10:30	Coal Science/Clean Coal and Gas to Fuels (Session 5)	Gasification Technologies (Session 6)
10:30 - 11:00		
11:00 - 11:30		
11:30 - 12:00		
12:00- 13:00	**Lunch**	
13:00 - 13:30	Carbon Management -3 (Session 7)	Clean Coal Demonstration and Commercial Projects (Session 8)
13:30 - 14:00		
14:00 - 14:30		
14:30 - 15:00		

	Wednesday (10/9/2024)	
	Track 1	**Track 2**
8:00 - 9:00	**Plenary Session III**	
9:00 - 10:00		
10:00 - 10:30	Clean Hydrogen (Session 9)	Coal Bed Methane and Shale Gas/Sustainability & Environment (Session 10)
10:30 - 11:00		
11:00 - 11:30		
11:30 - 12:00		
12:00- 13:00	**Lunch**	
13:00 - 13:30	Rare Earth Elements/Energy Storage (Session 11)	Combustion Technologies (Session 12)
13:30 - 14:00		
14:00 - 14:30		
14:30 - 15:00		

Times listed are Eastern Daylight Savings Time USA (Local Pittsburgh, PA, USA)

ADVISORY BOARD & INTERNATIONAL VICE CHAIRS

ADVISORY BOARD

Dr. Massood Ramezan, Chair, KeyLogic Systems, Inc., USA

Prof. Richard Bajura, West Virginia University, USA
Prof. Tarunjit Butalia, The Ohio State University, USA
Dr. Daniel Connell, CONSOL Energy Inc., USA
Mr. John Duddy, HTI, USA
Dr. Evan Granite, U.S. DOE/NETL, USA
Prof. Atsushi Ishihara, MIE University, JAPAN
Mr. Andrew Kramer, Gas Technology Institute, USA
Dr. Holly Krutka, University of Wyoming, USA
Mr. Francis Lau, EverBright & Associates, USA
Dr. Wenhua Li, NICE, CHINA
Dr. Rolf Maurer, Hamilton Maurer International, USA
Prof. Badie I. Morsi, University of Pittsburgh, USA
Prof. Masakatsu Nomura*, Osaka University, JAPAN
Dr. Alberto Pettinau, Sotacarbo S.p.A., ITALY
Dr. Andreas Richter, Freiberg University of Mining & Technology, GERMANY
Dr. Daniel Roberts, CSIRO, AUSTRALIA
Dr. Leslie Ruppert, U.S. Geological Survey, USA
Mr. Thomas Sarkus, U.S. DOE/NETL, USA
Mr. Brian Shaffer, U.S. Geological Survey, USA
Dr. Aleksander Sobolewski, Institute for Chemical Processing of Coal, POLAND
Mr. Don Stevenson, GTI Energy, USA
Mr. Gary Stiegel, Consultant, USA
Dr. Johan van Dyk, GTI, USA
Prof. Ting Wang, University of New Orleans, USA
Mr. Richard Winschel, Longbridge Energy Consulting, USA
*Emeritus

INTERNATIONAL VICE CHAIRS

Prof. Atsushi Ishihara, MIE University, JAPAN
Prof. Chung-Hwan Jeon (Steve Jeon), Pusan National University, SOUTH KOREA
Dr. Wenhua Li, CHN Energy, CHINA
Dr. Daniel Roberts, CSIRO, AUSTRALIA
Prof. Aleksander Sobolewski, Institute for Chemical Processing of Coal, POLAND
Dr. Johan van Dyk, GTI, USA

2024 SESSION CHAIRS

Prof. Richard Bajura, West Virginia University, USA
Dr. Omar Basha, ONB Engineering Research and Technical Services LLC, USA
Dr. Evan Granite, U.S. DOE/NETL, USA
Prof. Bingyun Li, West Virginia University, USA
Dr. Massood Ramezan, KeyLogic Systems, Inc., USA
Mr. Thomas Sarkus, U.S. DOE/NETL, USA
Mr. Richard Winschel, Longbridge Energy Consulting, USA

ANNOUNCING

Forty-second ANNUAL INTERNATIONAL PITTSBURGH Energy and Carbon Management CONFERENCE

Hosted by:

University of Pittsburgh

Location: TBA

Date: TBA

Abstracts must be submitted by March 1, 2025. Please submit an abstract including title, topic area, authors, affiliations, and contact information via email to:

ipcc@pitt.edu

PCC WEBSITE: www.pccpitt.org

ONB ENGINEERING RESEARCH AND TECHNICAL SERVICES

Highlighting Critical Barriers in the CCUS Value Chain to Develop technically and Economically Feasible Solutions

Omar M. Basha, Ph.D., P.E.

41st Annual International Pittsburgh Energy and Carbon Management Conference (PECMC)
Oct, 2024

CO_2 Industrial Emissions in the US

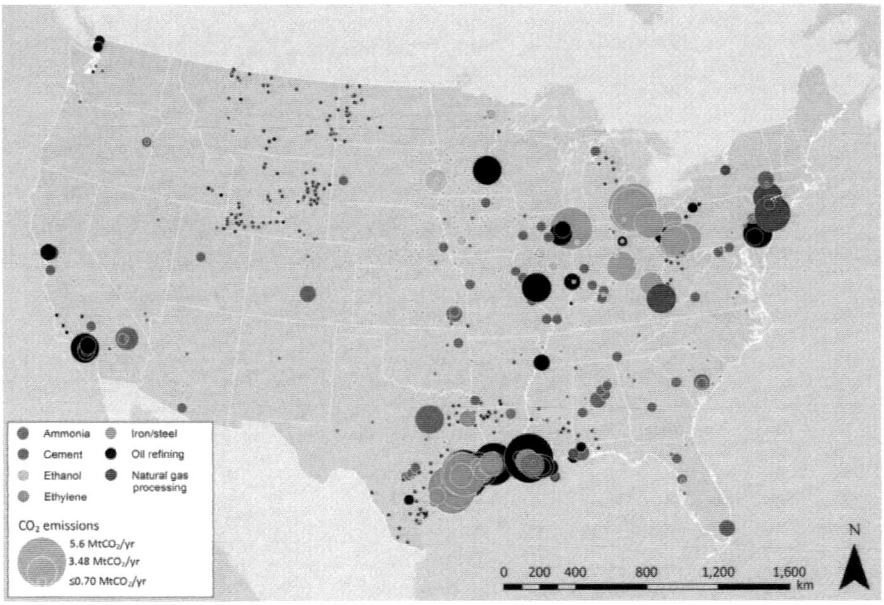

Industrial CO_2 sources distribution

- Gas processing has many micro-clusters across large states like Montana, Wyoming, Texas and Louisiana. Each hub serves multiple companies and 1000s of wells
- Ethanol is concentrated in the agricultural corn belt of Minnesota, South Dakota, Iowa and Nebraska with each hub serving the immediate farming communities
- Cement plants are across the country close to growing population centers
- Petrochemicals, such as ethylene, are mostly in Texas and Louisiana
- Oil refining is mostly on the Gulf Coast, with 6 other major hubs on the coasts or Mississippi river
- Iron/Steel in Michigan, Ohio, Pennsylvania, Indiana and Illinois near coal supplies
- Natural gas process in the large Marcellus reserves running through Pennsylvania

Major industrial point sources are regionally constrained due to resource or logistics drivers

2

Current Status of CO_2 Industry in USA

CO_2 pipeline current and planned status in USA

- **USA has the largest CO_2 pipeline network in the world** - ~5,150 miles (8,300km).
  - Several thousand additional miles are being added by 2025 to manage CO_2 from corn ethanol plants.
  - This network is small compared to oil and gas pipelines.
- **There are four key CO_2 pipeline hubs: Rocky Mountains, Permian Basin, Gulf Coast and the Mid-West.**
  - Significant existing pipeline is clustered around the Permian Basin oilfields of west Texas, where CO_2 is injected for enhanced oil recovery (EOR) to extract oil from ageing and difficult fields.
- **New CO_2 pipeline projects** have been announced, primarily in the Rocky Mountains and Midwest region.
  - The government aims to encourage investment through tax incentives and low-interest federal loans for expanding CO_2 pipeline capacity.
- **USA needs 68,000 miles of new CO_2 networks by 2050** to meet 1.5 °C emissions goals with an investment of $230 billion.
- **A major hindrance in CO_2 pipeline capacity addition will be the 'anti-pipe movement' in the US, which has derailed a number of oil and gas projects in the past and poses similar threats to new, large-scale CO_2 pipelines.**
  - The long pipeline approval interval, which can be as long as five years, may also deter investments.

Several sequestration sites in proximity to industrial carbon emission sources makes USA an attractive market for CCUS growth

Stages of CCUS Value Chain

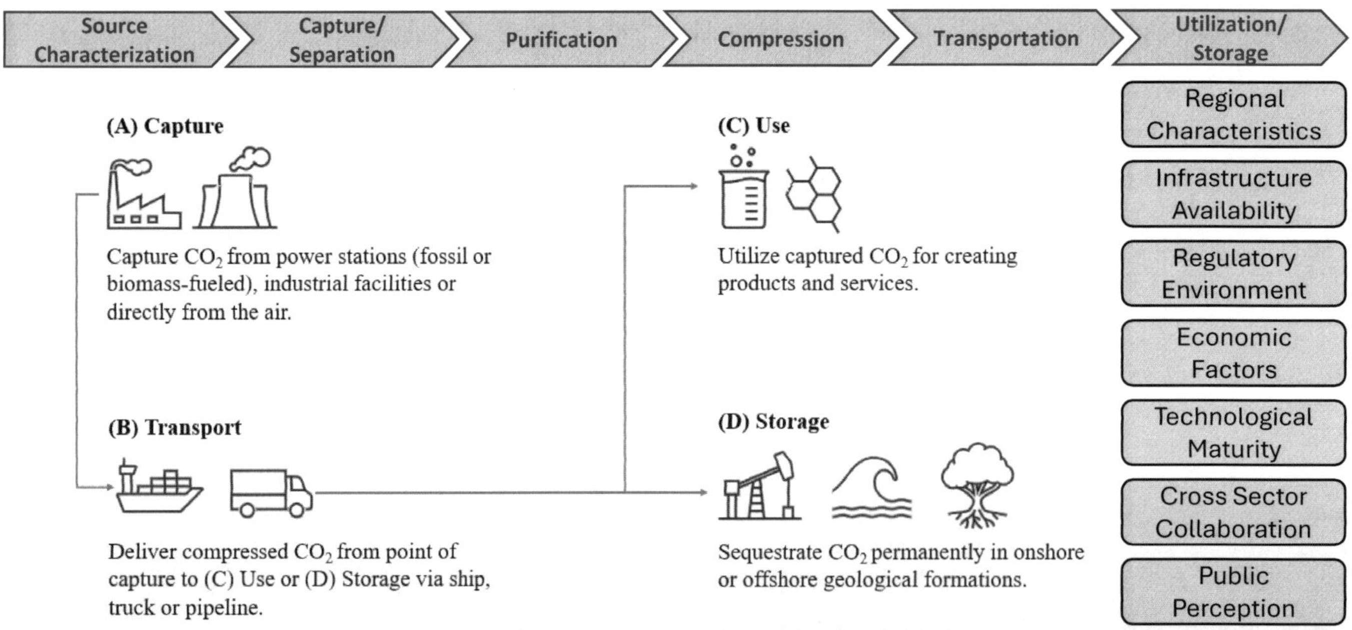

Stages of CCUS Value Chain

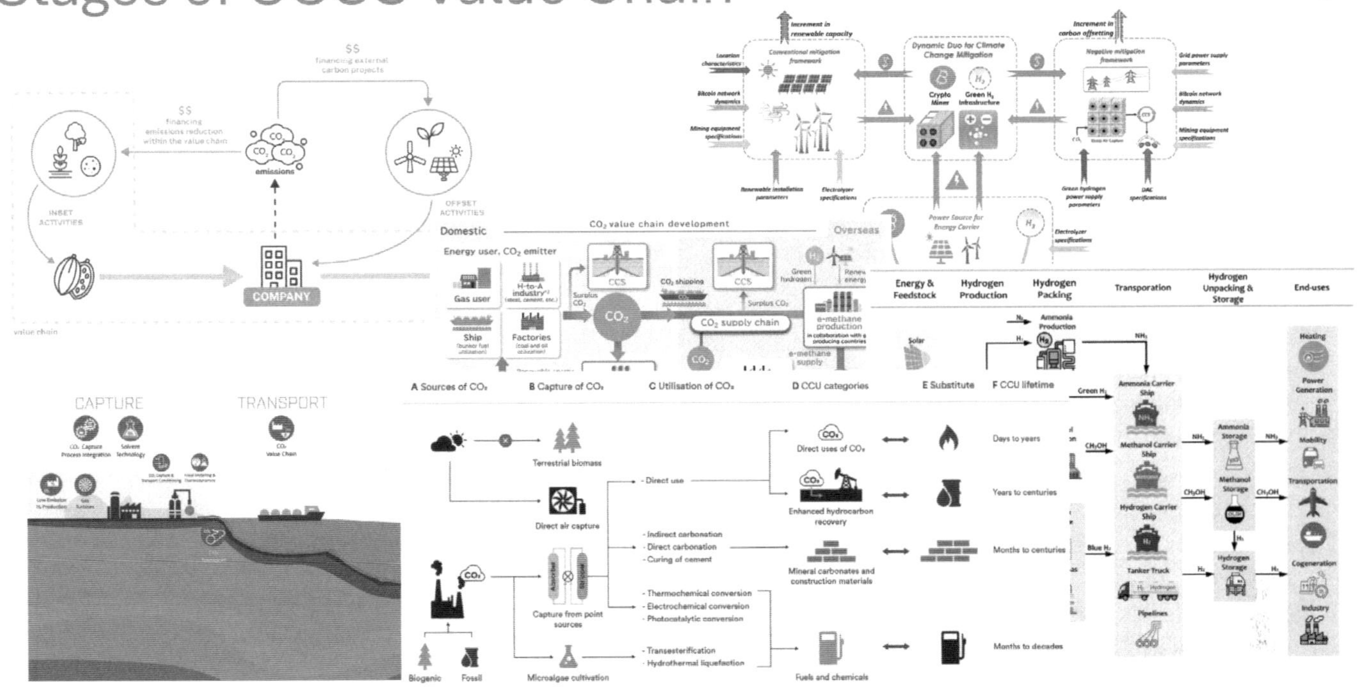

5

High-Level Technical Screening of CCUS Value Chains

Industry Clusters	Transport Infrastructure	Storage Sites

Mapping of emitters
- Emissions sources
- Amounts
- Composition
- Location
- Type of Industry

Utilization
- Technology options
- Renewable energy

Infrastructure Screening
- Existing pipelines
- Transport corridors
- Distance
- Transport capacity
- Waterways and ports
- Railways
- Roads

Storage Screening
- Saline aquifers or depleted reservoirs
- Onshore or offshore
- Storage capacity
- Injectivity
- Caprock properties
- Legacy wells
- Seismicity
- Protected areas
- Storage readiness level

6

Technical Characteristics of a CCU Value Chain –

Carbon Source Characterization

CO_2 Purity and Source Classification:

- No standard classification for CO_2 purity across industries.
- Various attempts at categorization (IPPC, Patricio et al., Zakkour, Cook, Jin et al.) lead to inconsistent boundaries.
- **CO_2 Purity Categories:**
  - High (>90% CO_2): Industrial processes such as fermentation, ethanol production, biogas.
  - Medium (50–90% CO_2): Syngas production (e.g., natural gas reforming).
  - Low (<50% CO_2): Combustion processes (power generation, cement production).

Carbon Capture

Challenges:

- High energy demand for low-purity sources.
- Need for purification for compatibility with transport and storage infrastructure.

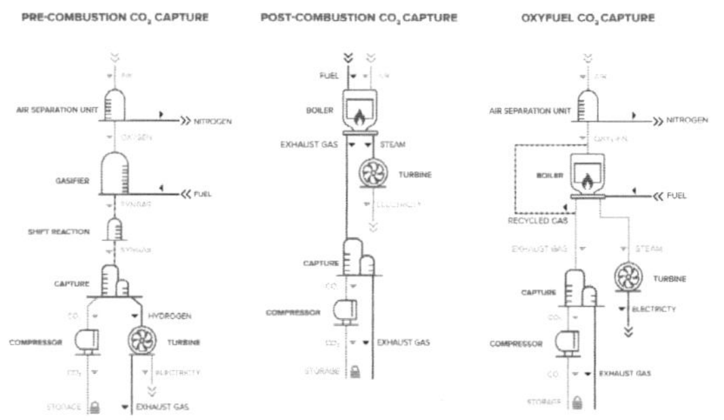

Carbon Purification

- CO_2 purification involves dehydration and compression.
- Impurities (H_2O, H_2S, SO_2, NO_2) raise pressure, lower pipeline capacity, and increase corrosion risk.

Purification Trade-off

- High-purity CO_2 prevents damage but adds costs.
- Skipping purification saves upfront costs but risks long-term damage.

7

Matching Sources with Separation Technologies

Source CO_2 Purity (increasing upward)

General Gas Conditioning

- Fermentation (100%)
- Food and Beverage (100%)
- Bioethanol (100%)
- Biogas production (99%)
- Refineries gas sweetening (96–99%)

Absorption with Physical Solvents

- Hydrogen production (100%)
- Ammonia production
- Haber-Bosch process (30–99%)
- EtO production (30–100%)
- Natural gas processing (2–70%)

Absorption with Chemical Solvents or Combustion using pure oxygen

- Lime kilns (24–32%)
- Cement kilns (22%)
- Oil refineries (8–24%)
- Blast furnace (22%)
- Iron and steel Industries
- Corex (30%)
- TGRBF (22–38%)
- Oil fired boilers (11–13%)

- Recovery boiler (13%)
- Pulp and Paper industry
- Energy production (13%)
- Power generation
- Coal fired boilers (11%)
- NG fired boilers (7–10%)
- Incineration of waste (10%)
- Glass industries (10%)
- Aluminium production (3–10%)
- Polyethylene production
- Cracker furnaces (5%)
- Gas turbines (3–4%)
- Carbon black mfg (2–5%)
- Brick dryers and kilns (1.5–4%)

IGCC (12–14%)

8

Technical Characteristics of a CCU Value Chain –

Carbon Utilization

CO_2 Utilization Categories:

- Direct Use: Food industry, metal industry, fire suppression, refrigeration.
- Mineral Carbonation: Produces aggregates for construction from mildly concentrated CO_2.
- Fuels Production: CO_2 to methanol, formic acid, synthetic natural gas.
- Chemicals Production: Organic acids, alcohols, urea, polymers (polycarbonates, polyurethanes).
- Enhanced Oil Recovery: High-volume CO_2 for oil extraction.

Challenges:

- **Compatibility:** Impurities in CO_2 streams can affect the process efficiency and cost of utilization.
- **Infrastructure:** Requires infrastructure to manage varying purity and flow rates efficiently.

CO_2 Use	Required Purity	TRL	Applications
Beverage Carbonation	>99%	9	Food and beverage industries
Renewable Methanol	>99%	7-8	Fuel production
Urea Production	>95%	9	Chemicals, fertilizers
Mineral Carbonation	Medium-Low	7-8	Producing aggregates for construction
Algae Cultivation	Medium-Low	4-7	Biofuels, chemicals

9

Economic Characteristics of CCU Value Chain

Overview of Economic Challenges:

- Most cost studies have focused on CCS, particularly carbon capture and transportation.
- Lack of Cost Studies for Utilization: There's limited data on the cost of retrofitting or building new plants to use recycled CO_2, a significant gap in the literature.

Carbon Capture Costs:

- Reported costs vary widely by sector and capture technology:
    - Iron and Steel Sector: \$40–65/tCO_2 (CCS), increasing to \$50–90/tCO_2 (CCU scenarios).
    - Cement Sector: Above \$65/tCO_2 avoided using post-combustion capture.
    - Petroleum Industry: Oxyfuel capture with costs at \$50–60/tCO_2.
- Capture Cost Ranges:
    - Power plants: \$30–150/tCO_2 avoided, depending on capacity, energy source, and technology.
    - Higher purity sources: \$10–41.6/tCO_2 avoided.

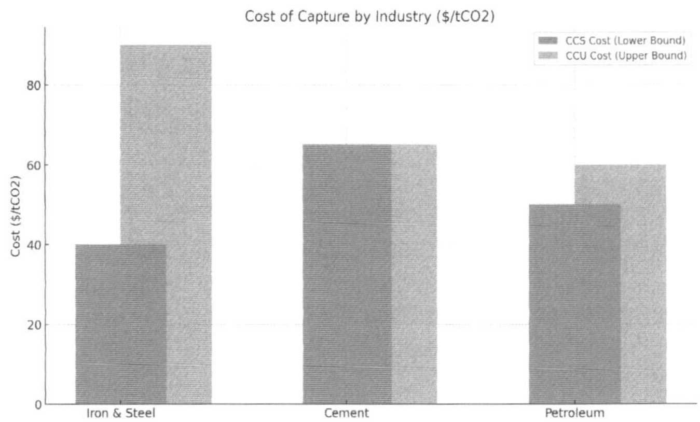

10

Economic Characteristics of CCU Value Chain

Transportation Costs:

- Pipeline Costs: \$1–5/tCO_2 for every 250 km, depending on flowrate.
- Ship Transportation: \$12–58/tCO_2, depending on distance and volume.
- Comparison:
  - Pipelines are more economical for short distances (<500 km) at around \$10–15/tCO_2 per 1000 km.
  - Ships become more economical for longer distances (>1500 km) at \$40/tCO_2 per 1000 km.

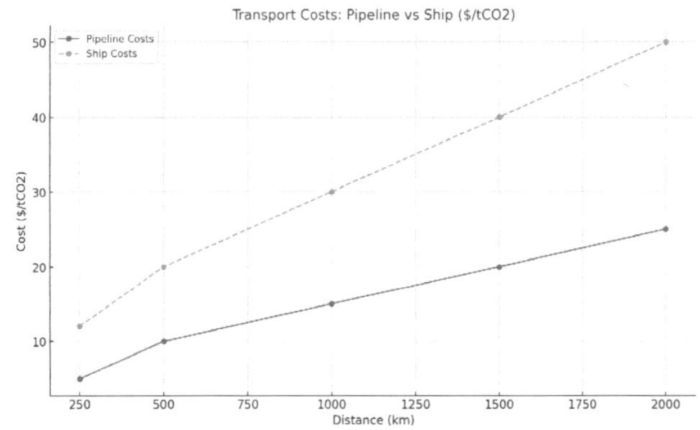

Cost Gaps for Utilization:

- Retrofitting or New Plant Installation: No significant data or estimates on the costs for industries to utilize captured CO_2.
- Challenges in Deployment: A combination of industrial plants and power plants may lead to more viable projects, but more data is needed.

Carbon value-chain - TRL

While individual components of the CCUS value chain, such as CO_2 utilization for synthetic fuels, are still emerging at lower TRL levels, the overall carbon capture, transport, and storage systems are well-developed and nearing maturity. This maturity presents an opportunity for significant short-term investments to accelerate deployment, particularly in regions with existing infrastructure.

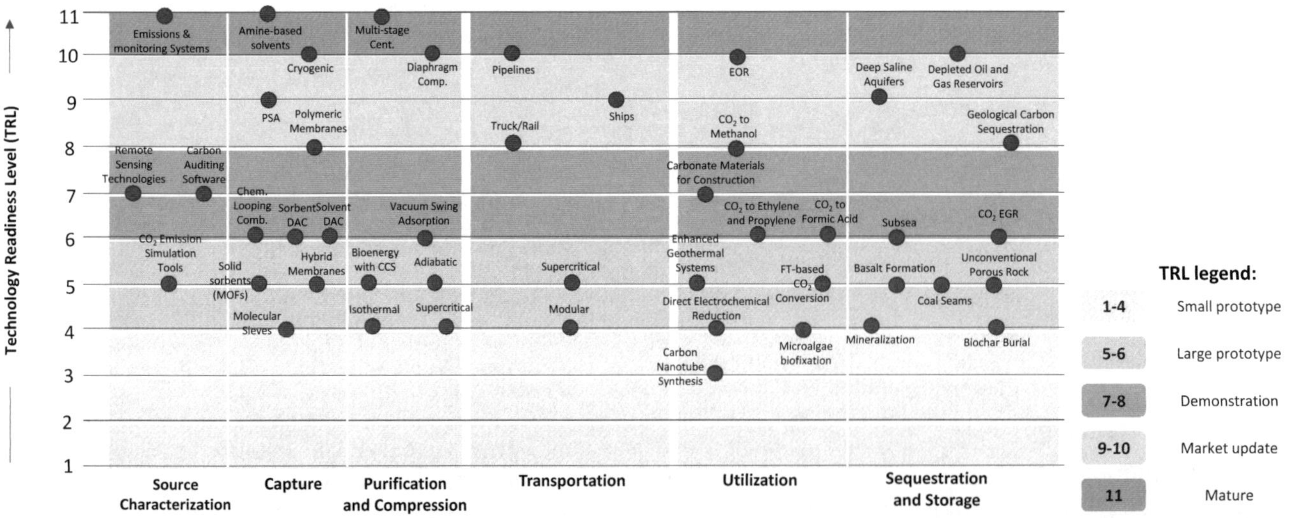

Hydrogen value-chain - TRL

Hydrogen use in synthetic hydrocarbons has the lowest TRL while low carbon hydrogen production is nearing mature TRL. Storage & distribution has attained mature TRL and is expected to be a key short-term investment area

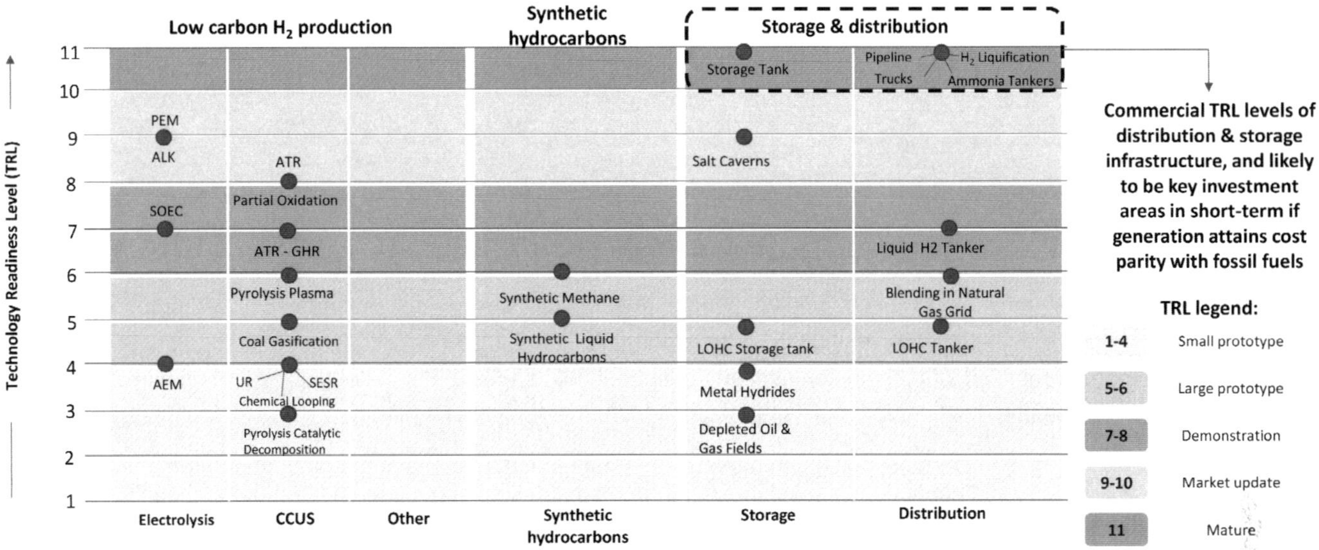

AEM: Anion exchange membrane, ATR = autothermal reformer, GHR: gas-heated reformer, LOHC: liquid organic hydrogen carrier, PEM: polymer electrolyte membrane, SOEC: solid oxide electrolyser cell, UR: Underground Reforming, SESR: Sorption enhanced steam Reforming

13

Carbon value-chain - Upstream LCOC

The levelized cost of capture (LCOC) for the upstream components of the Carbon value chain is shown across various industries.

The levelized cost of capture (LCOC) is highest for CO_2 capture, particularly for Direct Air Capture (DAC).

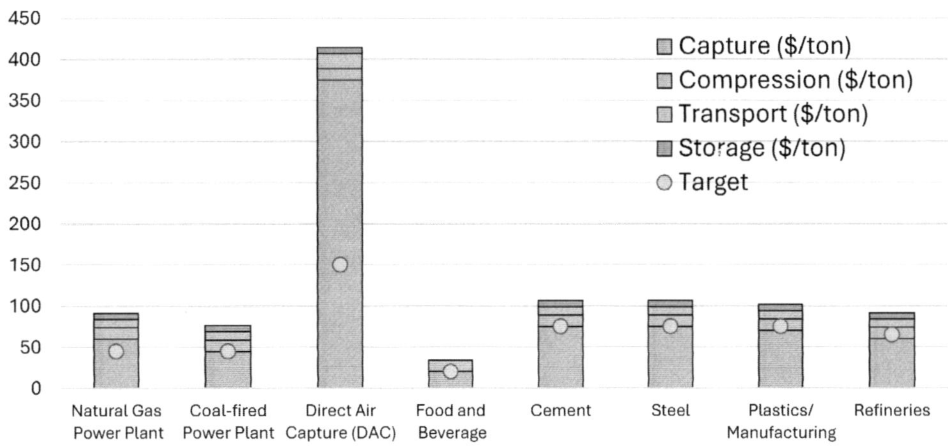

This is partially due to the variability in CO_2 purity across industries, leading to technology mismatches and higher costs. Aligning the right capture technologies with CO_2 sources can significantly reduce these costs."

14

Carbon value-chain - Downstream LCOC

Currently, only EOR meets LCOC targets, while CO_2 utilization technologies like CO_2-to-chemicals and fuels remain costly due to higher risk factors.

These costs are expected to decrease with further piloting and de-risking.

CO_2-to-building materials is approaching its target, while biological utilization and CO_2-to-synthetic fuels remain the farthest from achieving cost goals.

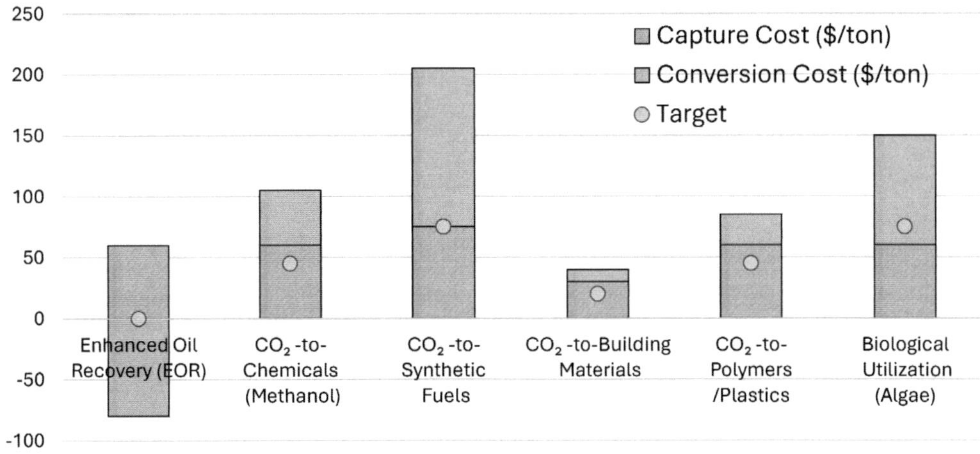

Cross-Sector Collaboration in CCUS – Unlocking Barriers

- Cross-sectoral collaboration between heavy industry, energy, and chemical sectors is essential for scaling up CCUS and overcoming challenges like infrastructure development, transport, and utilization.
- Each sector plays a critical role in creating an integrated value chain.
- Collaboration on standards, infrastructure investment, and policy advocacy can accelerate deployment.

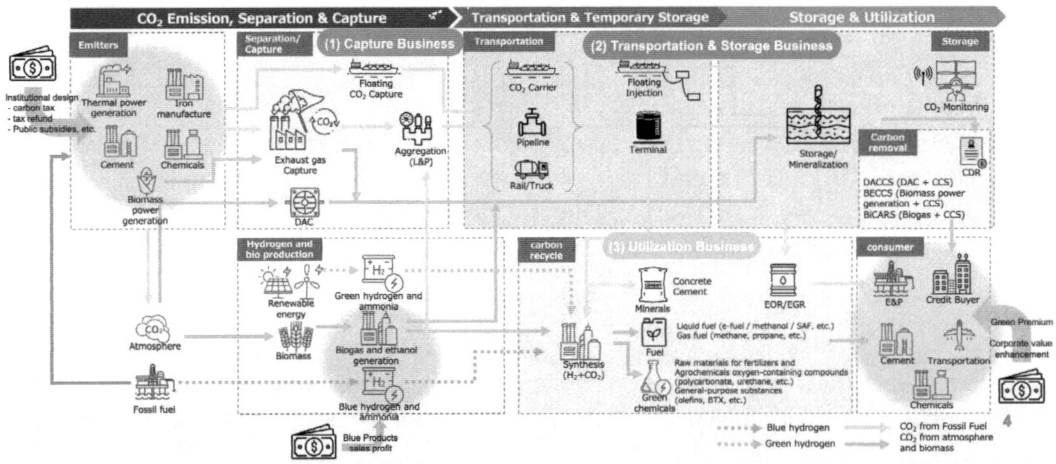

What are the Barriers in the CO_2 value chain?

5 key barriers:

Compatibility of Feed Streams:

- Different CO_2 capture processes result in varying levels of impurities.
- Impact: Increased conditioning and energy costs if not harmonized.

Disconnected Industries and Regulations:

- Cross-national regulatory differences hinder the global CCUS value chain.
- Impact: Delays in global-scale CCUS deployment.

Lack of CO_2 Quality and Purity Standards:

- No global standard for downstream CO_2 storage and conversion exists.
- Impact: Inconsistent CO_2 conditioning requirements across regions.

Insufficient Transport Infrastructure:

- The lack of transport infrastructure limits CCUS implementation.
- Impact: Restricted scalability of CCUS.

Limited Global Storage Capacity:

- The world needs expanded storage capacity through reservoir innovation.
- Impact: Limited ability to store CO_2 long-term.

Barrier 1 – CO_2 Compatibility Across Capture, Transport, and Storage

- CO_2 purity when captured can range from 75 to 99.9 vol.% - Common impurities are NO_X, H_2, Solvent degradation products.
- Table shows impurity levels from 5 CO_2 storage and transport stakeholders in Europe and US.
- Consequences:
  - Increased energy requirement for CO_2 handling
  - Higher operational cost
  - Responsibility of conditioning is unclear, and no CO_2 quality standards exist.

Species	A	B	C	D	E
CO_2 (mol%)	≥98	≥95%	–	–	≥97%
H_2O (ppm)	≤10	<480	≤30	≤30	≤50
H_2 (ppm)	≤600	≤1000	≤50	≤50	≤100
N_2 (mol%)	≤0.32	<4	–	–	<2
Ar (ppm)	≤500	<1	–	–	<2
CH_4 (mol%)	≤0.91	–	–	–	<0.5
CO (ppm)	≤300	<4250	≤100	≤100	≤100
O_2 (mol)	≤10	<10	≤10	≤10	≤10
H_2S (ppm)	≤9	≤20	≤9	≤9	≤20
SO_x (ppm)	≤10	<1	≤10	≤10	≤10
NO_x (ppm)	≤10	<1	≤10	≤10	≤10
Ammonia (ppm)	–	<50	≤10	≤10	–
Amine (ppm)	–	<1	≤10	≤10	–
Formaldehyde (ppm)	–	–	≤20	≤20	–
Hg (ppm)	≤0.03	≤0.03	≤0.03	≤0.03	≤0.03
Arsenic (ppm)	–	≤0.15	–	–	≤0.10
Lead (ppm)	–	≤0.2	–	–	≤0.10
Chromium (ppm)	–	≤0.5	–	–	≤0.3
Copper (ppm)	–	≤0.5	–	–	≤0.3
Nickel (ppm)	–	≤0.5	–	–	≤0.5
Zinc (ppm)	–	≤0.5	–	–	
dioxins, furans (ppm)	–	≤1.5	–	–	–
Glycol (ppm)	–	<40	–	–	–
Sulfur (ppm)	–	<35	–	–	–
C2 (mol)	≤2	–	≤3.5	–	–

Barrier 2 – The Need for Flexible CO_2 Transport Infrastructure

Transport Challenges:
- For volumes **<5 million tons/year** and distances **>300 km**, shipping is more cost-effective vs offshore pipelines.
- **Pipeline construction**: Takes 5-10 years and requires upfront investments.
- **Shipping**: Can scale faster (1.5-2 years to construct CO_2 carriers) and is currently the preferred option for stakeholders outside the US.

U.S. Context:
- In the U.S., pipelines are dominant form of CO_2 transport for EOR and CCUS projects.
- Long-distance pipeline projects face regulatory hurdles and public opposition, leading to delays and cost overruns.
- CO_2 shipping is not widely adopted, but with coastal storage sites and potential for offshore storage, could become a key part of future U.S. infrastructure.

Transport Method	Capacity	Cost	Scalability	Time to Construct	Flexibility
Pipelines	High (Million tons/yr)	High upfront (long-term savings)	Low (takes time to build new ones)	5-10 years	Low
Rail	Moderate (Medium scale)	Moderate (fuel, operation costs)	Moderate (existing infrastructure)	N/A (uses existing tracks)	Moderate
Truck	Low (Small scale)	High per unit transported	High (readily available trucks)	Immediate	High
Shipping	Moderate (Large-scale)	Moderate (cheaper than pipelines)	High (quick to scale)	1.5-2 years (build ships)	High

Existing and Future Carbon Infra. Mid-West becoming the newest CO_2 pipeline network hub

USA has 3 key CO_2 pipeline hubs, 4th hub is planned in the mid-west region

1: Rocky Mountains 2: Permian Basin 3: Gulf Coast 4: Mid-West (planned)

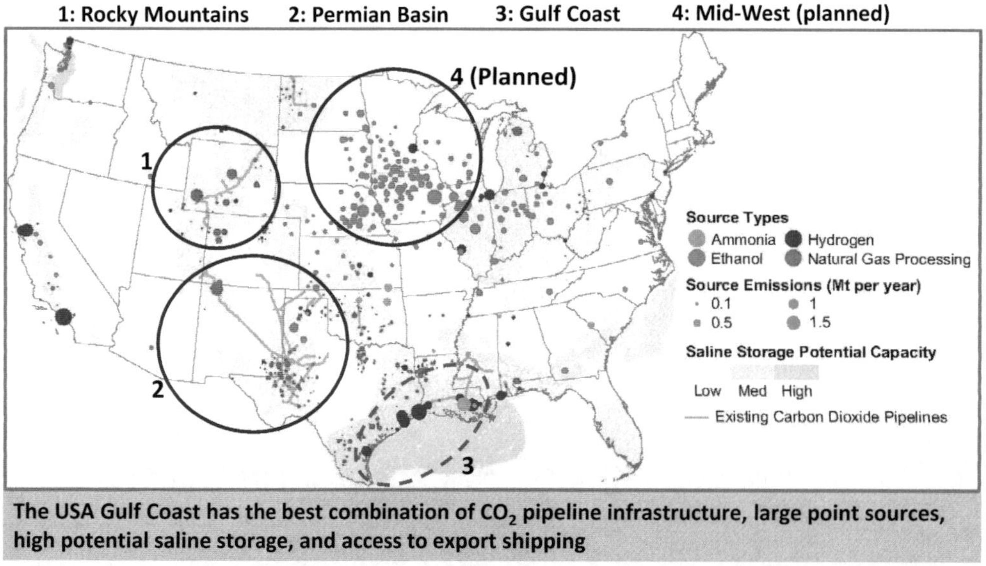

The USA Gulf Coast has the best combination of CO_2 pipeline infrastructure, large point sources, high potential saline storage, and access to export shipping

Barrier 3 – Upgrading Infrastructure for CO_2 Storage Capacity

Storage Challenges:

- The world needs 5000 Mtpa of storage capacity by 2050, but only 40 Mtpa is available today.
- U.S. geologic formations provide vast storage potential, but current infrastructure is a bottleneck.
- **Upgrading injection, transport, and monitoring systems in saline aquifers and offshore reservoirs will be essential to scale storage.**
- Federal investment and incentives are needed to make storage capacity upgrades economically viable.

U.S.-Regional Insights:

- The **Gulf Coast and Midwest** are crucial hubs for CO_2 storage, but they need rapid infrastructure upgrades.
- **Basaltic rock formations**, such as those found in the Pacific Northwest, may offer additional long-term storage potential but remain underexplored.

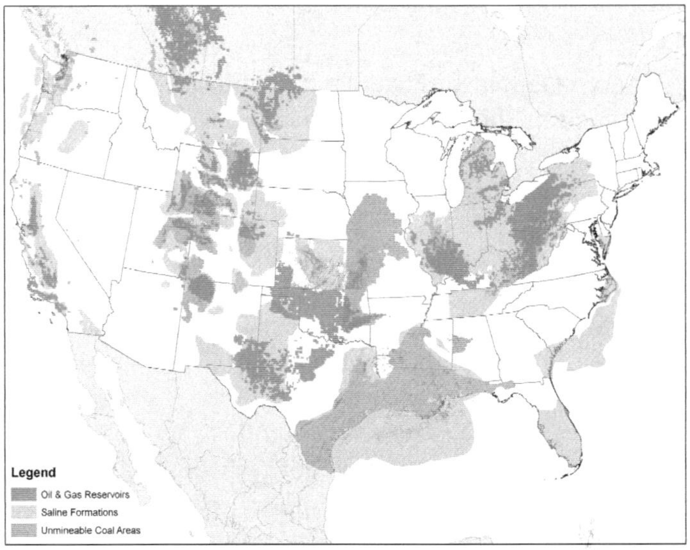

Barrier 4 – Standardized CO_2 Purities for Safe Storage and Effective Utilization - 1

CO_2 purity impacts both storage safety and market value for utilization:

- Impurities (e.g., NOx, H_2S) can cause corrosion and increase risk of leakage in wells used for long-term storage.
- Higher-purity CO_2 is essential for industrial processes such as methanol production and food-grade applications.

Wells must be retrofitted to handle impurities:

- Without proper global standards, wells must be adapted to store CO_2 with varying impurity levels, leading to higher costs and uncertainty.

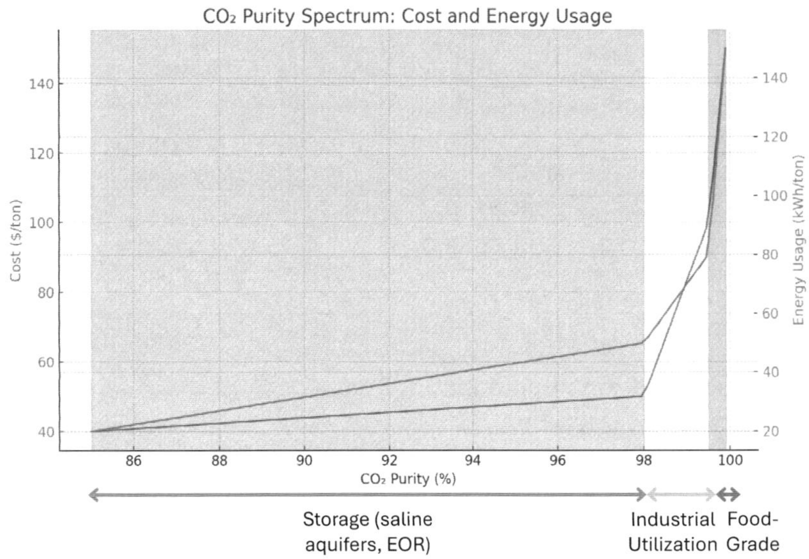

Barrier 4 – Standardized CO_2 Purities for Safe Storage and Effective Utilization - 3

A tiered CO_2 purity system is needed:

- Standards should distinguish between low-purity CO_2 for storage and high-purity CO_2 for utilization to balance costs and infrastructure needs.

U.S. Focus:

- In the U.S., 45Q tax credits incentivize CO_2 storage and utilization, but lack of national CO_2 purity standards creates uncertainty.
- The U.S. must develop flexible infrastructure to handle CO_2 with varying impurity levels, supporting long-term storage and high-purity utilization.

Carbon capture tax credit would increase under Inflation Reduction Act ($/tonne)

	Current	Inflation Reduction Act: POINT SOURCE	Inflation Reduction Act: DIRECT AIR CAPTURE
UNDERGROUND STORAGE (CO2)	$50	$85	$180
UTILIZATION (CO2)	$35	$60	$130
UTILIZATION IN ENHANCED OIL RECOVERY (CO2)	$35	$60	$130

As of July 28, 2022.
Sources: Clean Air Task Force; S&P Global Commodity Insights

23

Barrier 5 – Connecting Industry Interfaces: Regulations and Market Forces

- Cross-border collaboration is essential for CO_2 transport and storage.
- The U.S. must establish onshore storage policies to drive down costs and expand CCUS capabilities.
- CO_2 purity standards must be established to Sox optimize storage and utilization pathways.

Region	CO_2 Storage Regulation	Cross-Border Collaboration	Infrastructure Readiness	CCS Market & Funding
United States	Developing, needs clearer national frameworks for storage.	Minimal, mostly regional.	Moderate, fragmented pipelines.	IRA supports CCUS but gaps remain in transport/storage.
European Union	Strong frameworks, cross-border allowances.	High: Partnerships, Northern Lights Project.	High: Cross-border pipelines and shipping.	Significant EU funding via Green Deal and Horizon 2020.
Canada	Advanced regulations, led by Alberta.	Moderate: Emerging partnerships with the U.S.	High: Strong regional infrastructure.	Government-led, e.g., Alberta Carbon Trunk Line.
Norway	Global leader in offshore storage regulation.	Very high: EU partnerships via Northern Lights.	Very high: Advanced offshore infrastructure.	Government-backed Longship project.
China	Regulations in development, focus on national projects.	Low: Primarily domestic focus.	Moderate: Developing, gaps remain.	Heavy investment in CCUS pilots, regulatory maturity growing.
Middle East	Progress in UAE and Saudi Arabia, national strategies.	Low: Focus on national projects.	Moderate: Developing around oil/gas fields.	Government-driven projects like UAE's Al Reyadah.
Japan	Developing regulations and strategies, limited storage.	Low: No significant cross-border initiatives.	Low: Infrastructure gaps remain.	CCUS funded by government programs, slow progress.

24

Lessons from Global CCUS Projects

- 80% of projects that seek to commercialize CCS technology have failed, particularly in the power generation sector.
- Major reasons for failure include high capital costs, technical challenges, and insufficient financial incentives.
- Successful projects often rely on monetizing CO_2 through enhanced oil recovery (EOR), which is controversial in terms of actual CO2 reduction.
- Only about 10% of power sector CCS projects have been completed, compared to higher rates in industrial and gas processing sectors.

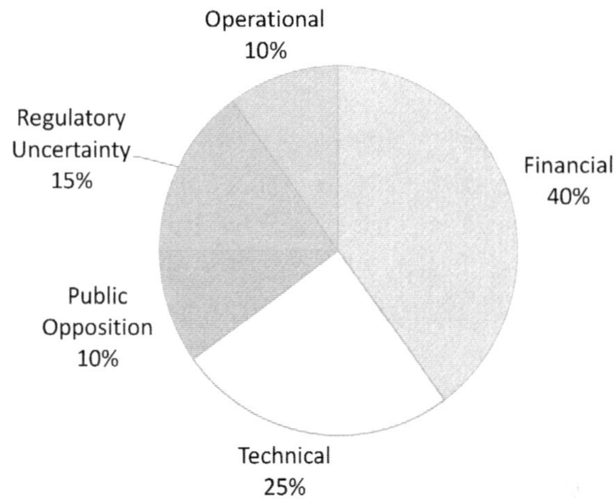

Global CCS Institute (2020). Global Status of CCS 2020. Global CCS Institute
International Energy Agency (IEA). CCUS in Clean Energy Transitions. IEA Report
Akhurst, M., et al. (2021). Storage Readiness Levels: Communicating the maturity of site technical understanding, permitting and planning. International Journal of Greenhouse Gas Control. DOI: 10.1016/j.ijggc.2021.103402
The Economist (2021). The challenges of deploying carbon capture and storage technologies. The Economist
MIT Energy Initiative (2016). Carbon Capture and Sequestration: Current state and future prospects. MIT Energy Initiative
Feenstra, C. F., Mikunda, T., & Brunsting, S. (2010). What happened in Barendrecht? Case study on the planned onshore carbon dioxide storage in the Netherlands. Barendrecht Case Study

25

CCS Challenges and Lessons from Failed Projects

1. Gorgon CCS Project (Australia)

- **Overview**: The Gorgon project aimed to capture CO_2 from the natural gas extraction process at the Gorgon LNG facility, storing it in an underground reservoir on Barrow Island.
- **Cost**: $3 billion
- **Challenge**: Despite its high cost, Gorgon captured only 30% of the intended CO_2, far below the target.
- **Technical Challenges**: The project faced significant injection issues due to unexpected complexities in the subsurface geology. These challenges made it difficult to inject CO_2 into the designated reservoir, leading to underperformance and higher operational costs

26

CCS Challenges and Lessons from Failed Projects

2. Petra Nova (Texas, USA)

- **Overview**: This project focused on capturing CO_2 from the W.A. Parish coal-fired power plant to use in enhanced oil recovery (EOR) operations.
- **Cost**: $1 billion, including $190 million in U.S. federal grants
- **Challenge**: Petra Nova faced operational challenges, with outages recorded on 367 days over three years. About 25% of these outages were due to issues with the CCS system itself.
- **Technical Challenges**: The project encountered significant problems with energy efficiency. The CCS system required additional power from a dedicated natural gas plant, which reduced the overall efficiency of the power generation process and led to higher operational costs than anticipated.

27

CCS Challenges and Lessons from Failed Projects

3. Enchant Energy (New Mexico, USA)

- **Overview**: This project was planned to retrofit the San Juan Generating Station (a coal-fired power plant) for CO_2 capture, with the goal of reducing emissions.
- **Cost:** Seeking $1 billion in loans
- **Challenge:** The project is significantly delayed and still burning coal without any CCS in place.
- **Technical Challenges:** Enchant Energy struggled with financing, as the high upfront capital costs, combined with difficulties in securing funding, have stalled the implementation of CCS technology. Additionally, the project faced technical challenges in integrating CCS into the existing coal-fired infrastructure.

28

Success Story: Quest – Industrial CCS in Canada

Overview:

- **Project**: Quest is Shell's pioneering CCS project, capturing CO_2 from its oil sands operations in Alberta, Canada.
- **Start Date**: Operational since 2015

Technical Highlights:

- CO_2 is captured from **hydrogen production** at the Scotford facility and stored underground.
- Robust monitoring systems ensure safe, long-term storage.

Key Achievements:

- Captures over 1 million tonnes of CO_2 annually.
- First industrial project to store such volumes, proving scalability.
- 30% cost reduction since operations began, showcasing efficiency.

Northern Lights – Cross-Border CO_2 Transport and Storage Networks

- **Overview**: A collaboration between **Norway, the EU, and industry partners** for CO_2 transport and storage.

Technical Highlights:

- CO_2 transported by ship to a central hub and stored under the North Sea.
- Capacity: 1.5 million tonnes/year, scalable to 5 million tonnes/year.

Key Achievements:

- First large-scale cross-border CO_2 storage network in Europe.
- Will store captured CO_2 from industries in multiple countries.

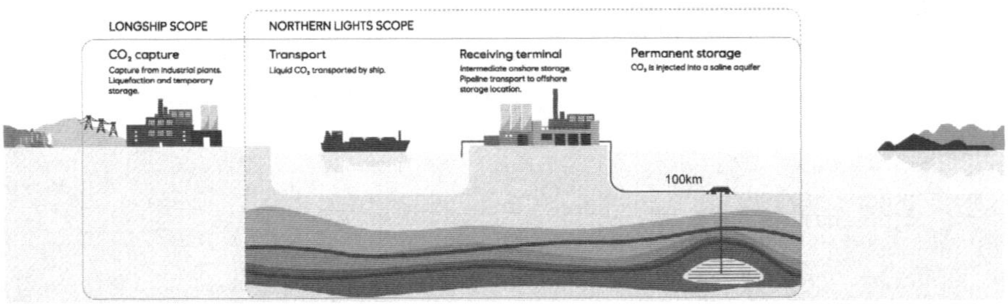

Climeworks – Direct Air Capture and CO_2 Utilization

- **Overview**: Climeworks captures CO_2 directly from the air using specialized filters.

Technical Highlights:
- Captures ~4,000 tonnes/year of CO_2.
- Uses renewable energy to power the process, minimizing its carbon footprint.

Key Achievements:
- World's first commercial-scale direct air capture plant.
- CO_2 is sold for commercial uses, such as synthetic fuels and carbonation for beverages.

31

Demand Roadmap for CO_2 through 2050 (1/3)

	Phase 1 2022 to 2025	Phase 2 2025 to 2030	Phase 3 2030 to 2035	Phase 4 2035 to 2050
Refining and petrochemicals	Pilot CO_2 capture for storage and enhanced oil recovery (EOR).	Expand CO_2 capture at refineries and explore utilizing CO_2 as a feedstock for chemicals.	Scale CO_2-based chemicals and polymers for large-scale industrial applications.	Achieve near-complete CO_2 capture across all refining and petrochemical processes.
New industrial applications	Pilot CO_2 utilization in concrete production and test initial CO_2-to-synthetic fuels technologies.	- Scale CO_2-based concrete production and deploy early-stage CO_2-to-methanol commercially. - Begin establishing CO_2 purity standards for utilization in synthetic fuels and building materials.	Full-scale deployment of CO_2-to-synthetic fuels for aviation and heavy industry.	Achieve 100% CO_2-based industrial processes, including materials and synthetic fuels.
Transportation	- Develop local CO_2 transportation networks using pipelines, trucks, and short-distance transport. - Explore onshore storage options to reduce costs.	- Expand transport networks to connect capture sites with sequestration hubs. - Ensure CO_2 purity standards are incorporated into infrastructure planning.	- Expand long-distance routes to support industrial demand. - Ensure infrastructure expansion aligns with CO_2 purity standards.	- Build a fully connected CO_2 pipeline network across regions and borders. - Full-scale adoption of CO_2-to-fuel aviation and marine vessels.
Power & energy storage	Conduct CO_2 capture trials at natural gas and coal-fired power plants.	Scale CO_2 capture systems for power plants, focusing on large-scale adoption.	Integrate CO_2 sequestration at all major power generation facilities.	Fully integrate CO_2-to-fuel plants and sequestration at scale across the power sector.
Export	Begin development of small-scale CO_2 export infrastructure for sequestration and industrial use.	Begin scaling CO_2 exports for storage and industrial use in Europe and Asia.	Scale global CO_2 export markets for industrial use and storage.	Expand global CO_2 export markets to meet industrial demand and achieve global sequestration goals.

32

Supply Roadmap for CO_2 through 2050 (2/3)

	Phase 1 2022 to 2025	Phase 2 2025 to 2030	Phase 3 2030 to 2035	Phase 4 2035 to 2050
CO_2 Capture Technologies	Pilot CO_2 capture at industrial plants (cement, steel, power). Launch direct air capture (DAC) pilots.	Expand industrial and energy CO_2 capture. Scale DAC and low-energy capture technologies.	Deploy DAC and BECCS at scale. Further reduce capture costs.	Near-complete CO_2 capture across all major industries.
Transport Infrastructure	Start local CO_2 pipeline and truck transport networks. Explore offshore shipping options.	Expand regional pipeline networks and CO_2 shipping infrastructure.	Expand long-distance pipelines and shipping for cross-border transport.	Fully connected CO_2 pipeline and marine transport networks.
Storage Capacity and Infrastructure	Pilot storage in saline aquifers and depleted oil fields. Begin EOR projects with CO_2 injection.	Increase onshore and offshore geological storage capacity. Scale enhanced oil recovery (EOR).	Scale offshore and onshore storage hubs with MMV systems.	Operate global-scale CO_2 storage hubs with geological capacity for millions of tons.
Utilization Pathways	Small-scale CO_2 utilization in concrete and synthetic fuels.	Scale CO_2-to-fuels and CO_2-based polymers for industry.	Full-scale deployment of CO_2-to-fuels and CO_2-to-chemicals.	Widespread use of CO_2-based products (fuels, chemicals, and building materials).
Emerging Technologies and Pathways	Test biomass-based CO_2 capture (BECCS) and ocean-based capture. Investigate early-stage CO_2 mineralization methods.	Pilot offshore CO_2 storage and advanced electrochemical CO_2 conversion. Continue developing CO_2 mineralization and biological utilization.	Test advanced ocean-based CO_2 capture, biological CO_2 utilization, and novel electrochemical pathways.	Commercialize advanced CO_2-to-fuel technologies, including biofuels and synthetic fuels. Scale CO_2 mineralization for permanent storage in construction materials.

Environmental Justice Roadmap for CO_2 through 2050 (3/3)

	Phase 1 2022 to 2025	Phase 2 2025 to 2030	Phase 3 2030 to 2035	Phase 4 2035 to 2050
Environmental justice	• Identify DACs most impacted by CO_2 emissions and understand how CCUS projects can benefit these communities. • Establish pollution reduction targets for areas surrounding major industrial and power plants where CO_2 capture is being implemented. • Increase air quality monitoring near CO_2 capture facilities in DACs. • Develop community outreach programs to ensure DACs are aware of CO_2 reduction efforts and how they can benefit.	• Guarantee that 40% of CCUS-related investments in pilot projects flow to DACs, aligning with federal justice goals (e.g., Justice40). • Ensure DACs near CO_2 capture and storage facilities see a reduction in pollution from adjacent industries. • Expand monitoring infrastructure in DACs to track air quality improvements and the progress of pollution reduction efforts.	• Reduce CO_2 pollution levels in DACs to align with state and national averages, focusing on regions impacted by industrial CO_2 emissions. • Integrate DACs into broader CCUS networks, creating opportunities for local economic development through job creation in CCUS projects.	• Continue reducing CO_2 pollution levels to meet EPA health standards and further improve air quality in DACs. • Ensure that DACs receive ongoing benefits from long-term CCUS infrastructure, including economic growth, improved health outcomes, and environmental remediation. • Promote community-led initiatives for local pollution control and environmental justice in the context of national and global CO_2 reduction efforts.

1. Use the White House's July 2021 "Interim Implementation Guidance for the Justice40 Initiative," which defines "disadvantaged" as: 1) low income, high and/or persistent poverty; 2) high unemployment and underemployment; 3) racial and ethnic residential segregation; 4) linguistic isolation; 5) high housing cost burden and substandard housing; 6) distressed neighborhoods; 7) high transportation cost burden and/or low transportation access; 8) disproportionate environmental stressor burden and high cumulative impacts; 9) limited water and sanitation access and affordability; 10) disproportionate impacts from climate change; 11) high energy cost burden and low energy access; 12) jobs lost through the energy transition; 13) limited access to healthcare

CCUS Value Chain Recommendations

- **Collaboration:** Involve all stakeholders, including academia and industry.
- **Data Transparency:** Push for open access to geological and emission data.
- **Hard-to-Abate Industries:** Focus on industries like cement and steel for emission capture.

35

Thank You!

www.onbengineering.com

info@onbengineering.com

36

International Pittsburgh Energy and Carbon Management Conference

41st Annual event

OCT. 9th, 2024
The Honorable Charles McConnell
Executive Director, Carbon Management and Energy Sustainability
Assistant Secretary of Energy - US DOE - (2011-13)

UH Energy
UNIVERSITY OF HOUSTON

Outline

1 **WHY CCUS**
- Energy Sustainability Requires Real Transformation
- CCUS is a Lynchpin technology
- CCUS is Not an Option – It is a Requirement!

2 **VIEW FROM MARKETS & ECONOMY**
- Infrastructure Must Support CCUS and Key Markets
- Hard to Abate Markets Decarbonization
- H2 Economy – far more than transportation
- Electricity grid – for growth and resilience

3 **PIPELINES**

4 **FOCUS MUST BE ON WHAT MATTERS – RELEVANCE AND IMPACT**

2

UH Energy
UNIVERSITY OF HOUSTON

Energy Sustainability Requires Real Transformation

UH Energy UNIVERSITY OF HOUSTON

Underlying the Energy Transition: Energy Trilemma

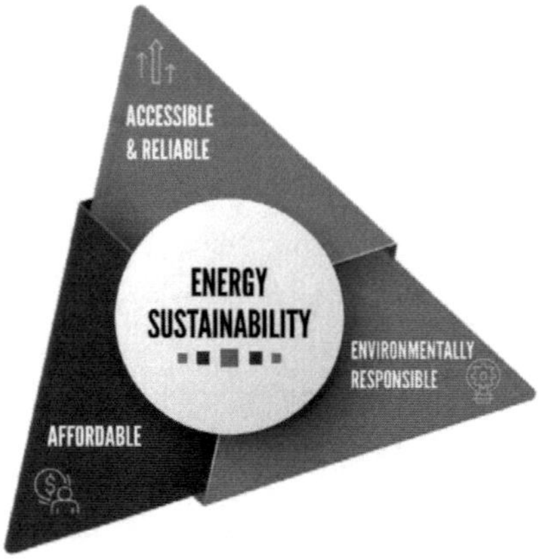

Global CO2 Emissions, 1900-Present

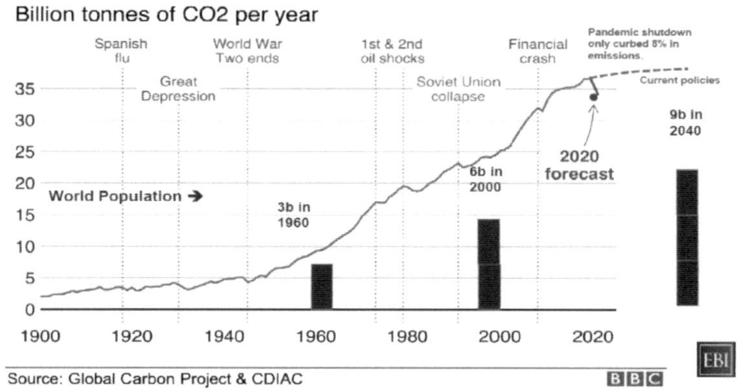

3

What is CCUS? Carbon Capture Utilization and Storage

UH Energy UNIVERSITY OF HOUSTON

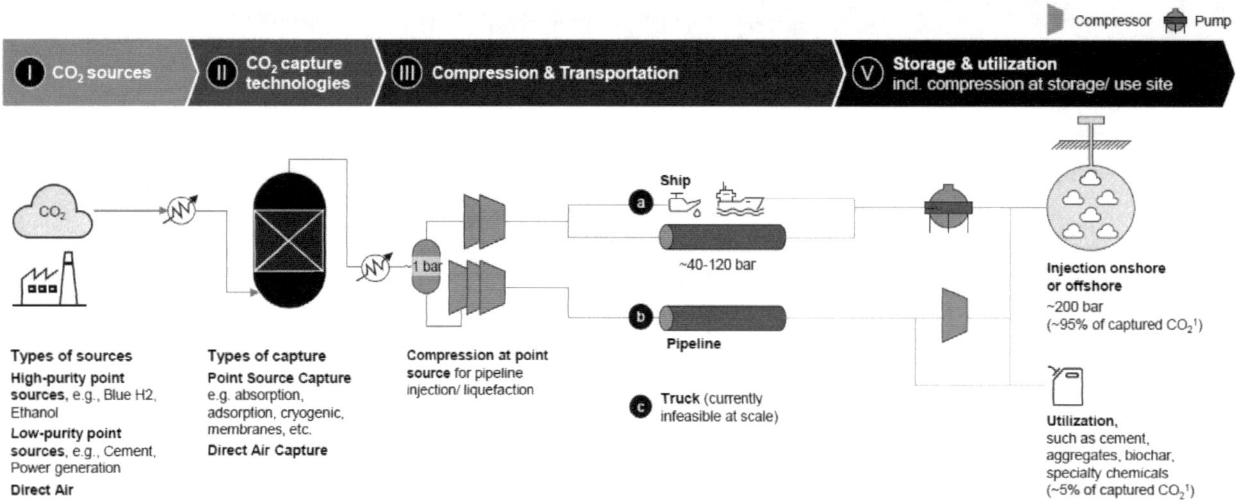

Successful execution of a CCUS project requires planning and collaboration across a range of technologies and capabilities, and may vary significantly depending on specific context

1. Projection for 2050, IEA SDS and Net Zero by 2050 scenarios

Source: McKinsey CCUS Cost Model: IEA

4

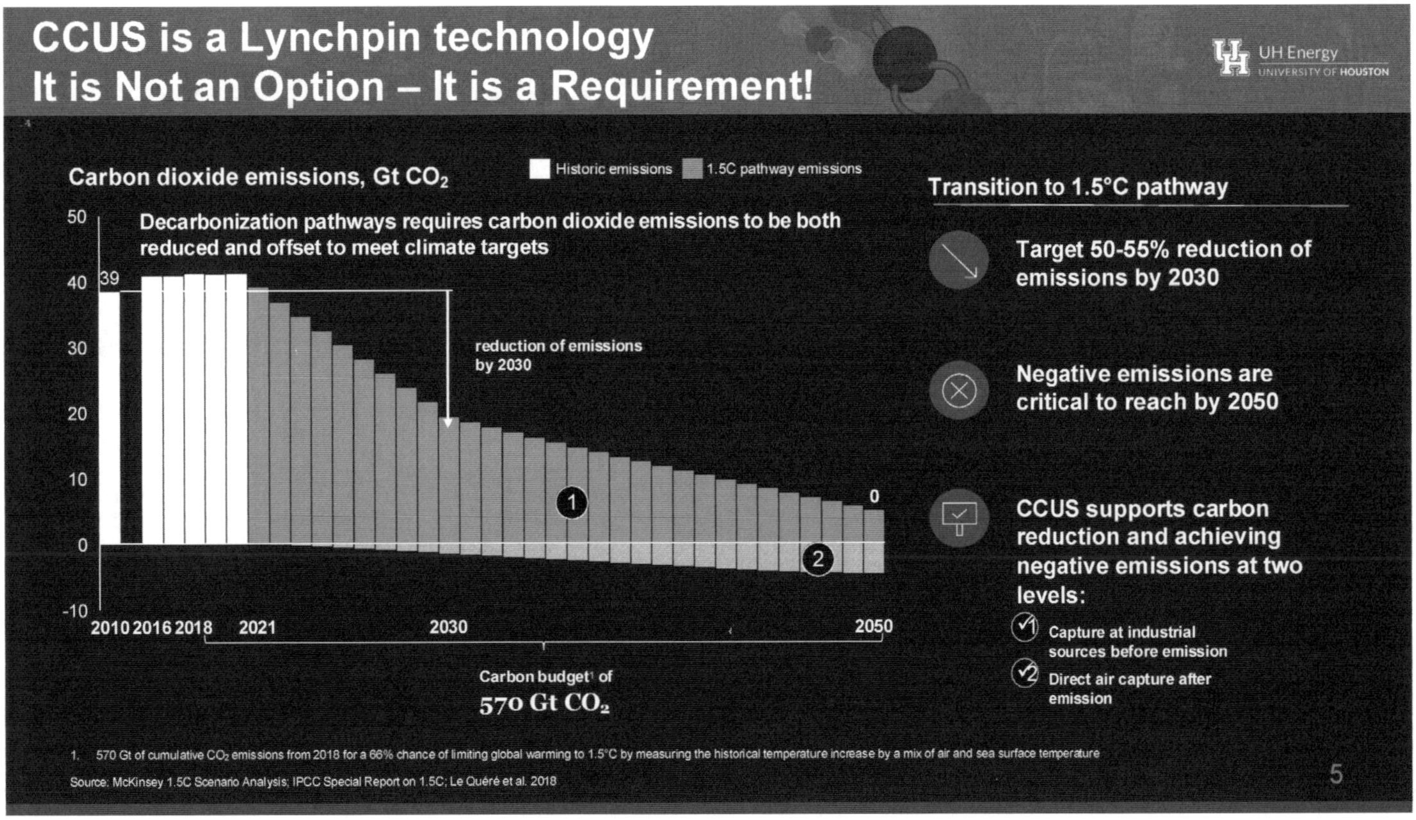
CCUS is a Lynchpin technology
It is Not an Option – It is a Requirement!
UH Energy
UNIVERSITY OF HOUSTON
Carbon dioxide emissions, Gt CO_2
Historic emissions
1.5C pathway emissions
Decarbonization pathways requires carbon dioxide emissions to be both reduced and offset to meet climate targets
50
40
30
20
10
0
-10
39
reduction of emissions by 2030
0
1
2
2010 2016 2018 2021 2030 2050
Carbon budget[1] of
570 Gt CO_2
Transition to 1.5°C pathway
Target 50-55% reduction of emissions by 2030
Negative emissions are critical to reach by 2050
CCUS supports carbon reduction and achieving negative emissions at two levels:
1 Capture at industrial sources before emission
2 Direct air capture after emission
1. 570 Gt of cumulative CO_2 emissions from 2018 for a 66% chance of limiting global warming to 1.5°C by measuring the historical temperature increase by a mix of air and sea surface temperature
Source: McKinsey 1.5C Scenario Analysis; IPCC Special Report on 1.5C; Le Quéré et al. 2018
5

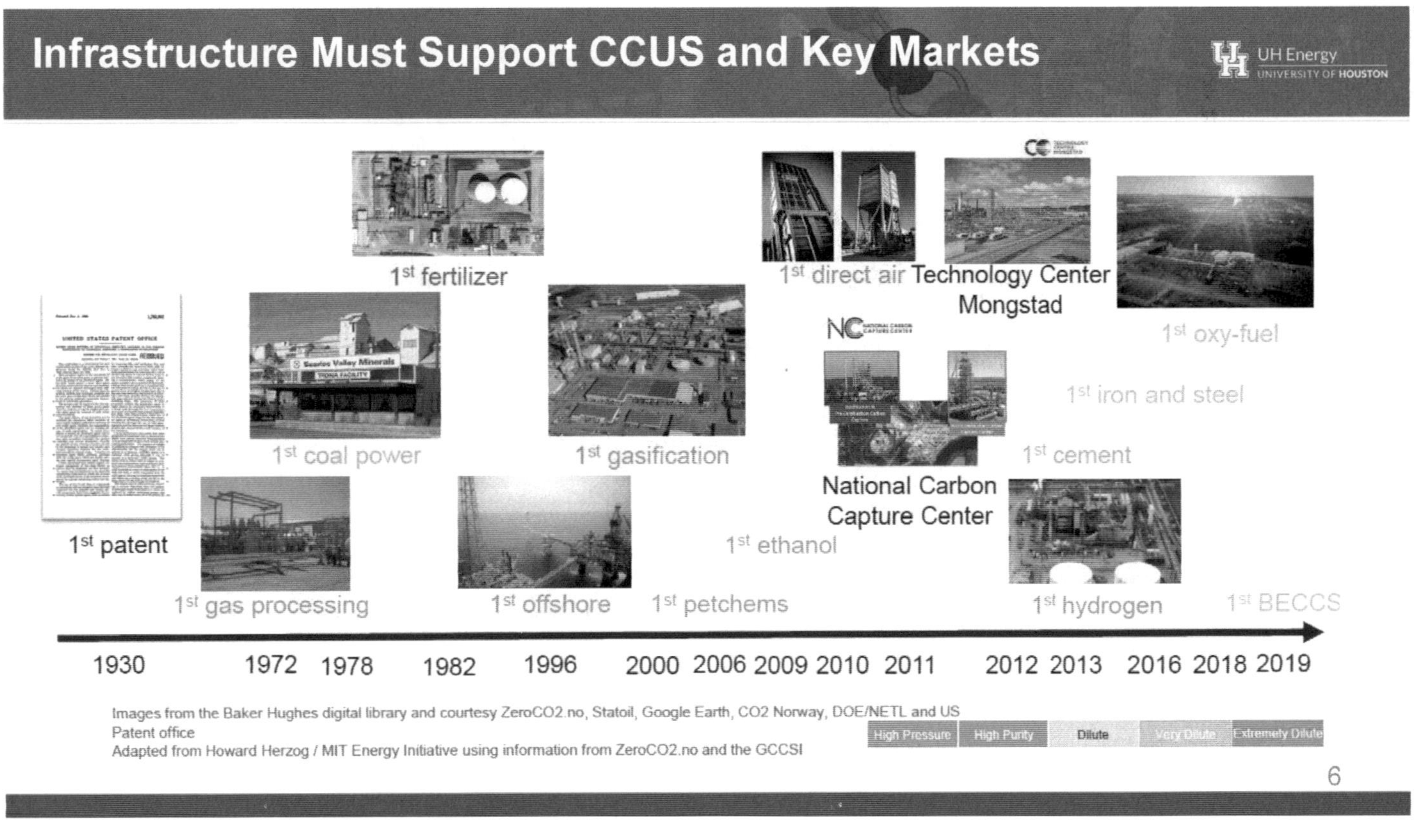
Infrastructure Must Support CCUS and Key Markets
UH Energy
UNIVERSITY OF HOUSTON
1st fertilizer
1st direct air
Technology Center Mongstad
1st oxy-fuel
1st patent
1st coal power
1st gasification
National Carbon Capture Center
1st iron and steel
1st cement
1st ethanol
1st gas processing
1st offshore
1st petchems
1st hydrogen
1st BECCS
1930 1972 1978 1982 1996 2000 2006 2009 2010 2011 2012 2013 2016 2018 2019
Images from the Baker Hughes digital library and courtesy ZeroCO2.no, Statoil, Google Earth, CO2 Norway, DOE/NETL and US Patent office
Adapted from Howard Herzog / MIT Energy Initiative using information from ZeroCO2.no and the GCCSI
High Pressure
High Purity
Dilute
Very Dilute
Extremely Dilute
6

Electricity grid – for growth and resilience

UH Energy
UNIVERSITY OF HOUSTON

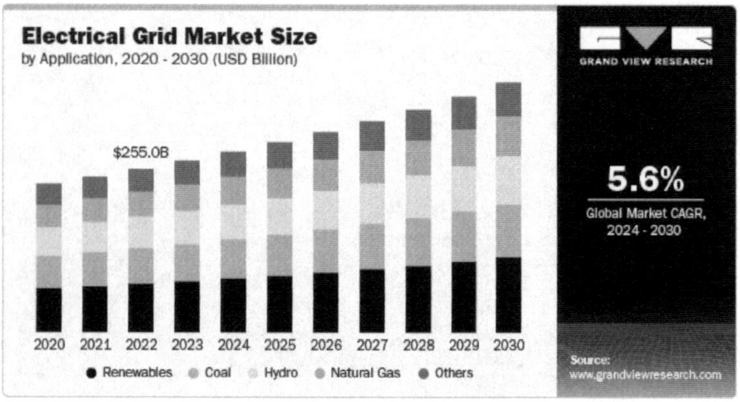

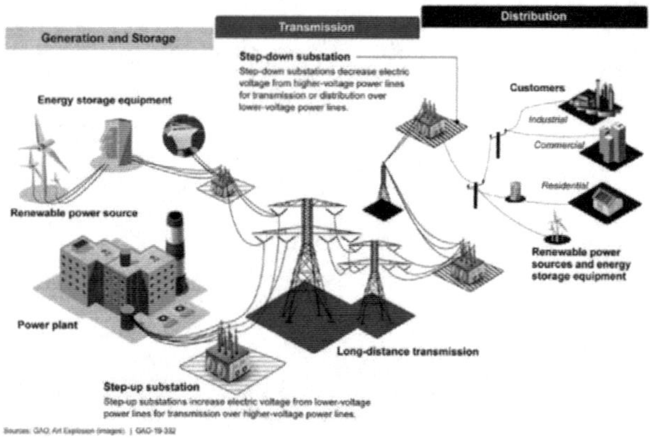

7

Pipelines

UH Energy
UNIVERSITY OF HOUSTON

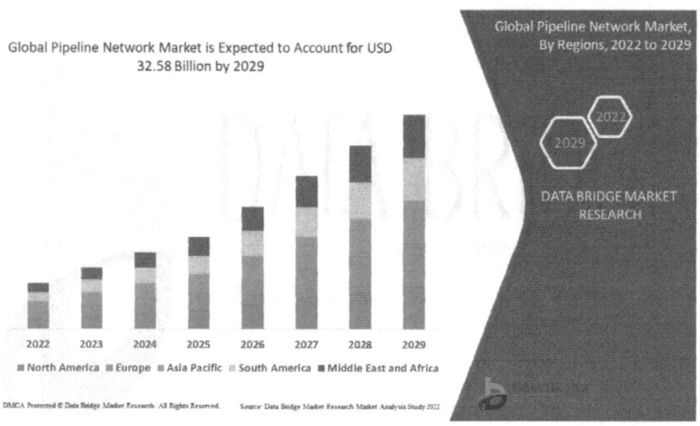

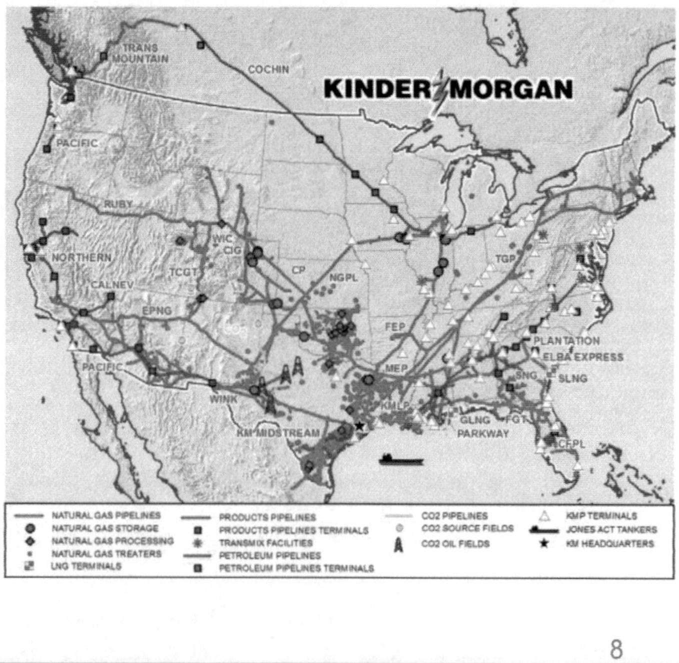

8

UH Energy
UNIVERSITY OF HOUSTON

FOCUS MUST BE ON WHAT MATTERS – RELEVANCE AND IMPACT

9

UH Energy
UNIVERSITY OF HOUSTON

Carbon Capture, Utilization and Storage - Lynchpin for the Energy Transition

https://uh.edu/uh-energy-innovation/uh-energy/energy-research/white-papers/white-paper-files/future-of-houston-updated.pdf

https://dualchallenge.npc.org/

CCUS Infrastructure: Preparing for the Future of Houston

https://uh.edu/uh-energy-innovation/uh-energy/energy-research/white-papers/white-paper-files/ccus-lynchpin-for-energy-transition-web1.pdf

NATIONAL PIPE LINE CONFERENCE
PIPE LINE CONTRACTORS ASSOCIATION

Thank you!

UNIVERSITY of HOUSTON
UH ENERGY

Center for Carbon Management in Energy
UNIVERSITY OF HOUSTON

Lessons Learned from Canadian CCUS Projects

2024

INTERNATIONAL CCS KNOWLEDGE CENTRE

Presented by:
Laura Westover, P.Eng.
Project Manager

Accelerating the path to net-zero emissions

The International CCS Knowledge Centre is leading the world to a sustainable future by sharing insights and expertise on carbon capture and storage (CCS) and other solutions to address climate change.

We are independent, trusted advisors with unparalleled experience developing CCS projects, fostering collaboration and the exchange of knowledge to cut greenhouse gas emissions and achieve global net-zero goals.

INTERNATIONAL CCS KNOWLEDGE CENTRE

Canadian CCS Implementation

INTERNATIONAL
CCS KNOWLEDGE
CENTRE

Canada – A Global CCS Leader

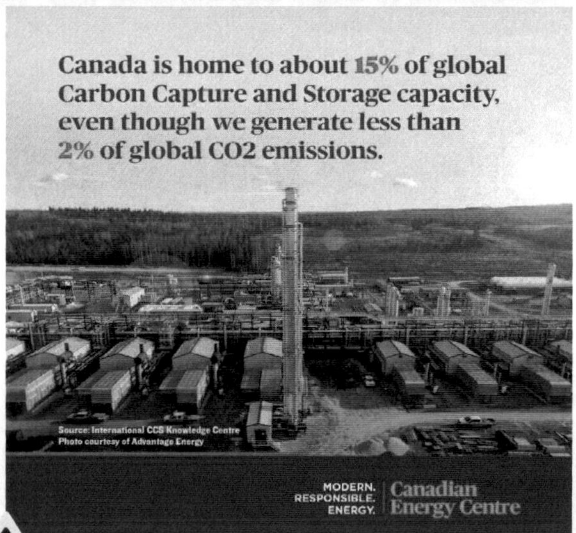

EXPERIENCE CAPTURING AND STORING CO_2 :

5 COMMERCIAL CAPTURE FACILITIES

- Boundary Dam Unit 3 CCS
- Quest
- Nutrien & NWR Sturgeon Refinery (CO_2 to ACTL, one group, two facilities)
- Advantage Energy's Glacier CCS

ALBERTA CARBON TRUNK LINE

- World's largest-capacity CO_2 pipeline, operated by Wolf Carbon Solutions
- 240 kms, 14.6 Mtpa capacity
- Currently running at ~10% capacity

WEYBURN-MIDALE CO_2 STORAGE

- >40 million tonnes since 2000
- Aquistore research project – 500,000+ tonnes

The BD3 ICCS Project Overview

- World's first post-combustion coal-fired CCS project **fully integrated** with a power station.
- Life extended the 45-year-old Boundary Dam Unit 3.
- Favored by economics at the time.
- Aided by **$240 CAD million-dollar federal grant. Total CAPEX: $1.5B ($900M for the Capture Plant)**
- Executed as a two-part project:
    - Power island upgrade
    - CCS retrofit
- **Capture operations began October 2014.**
- CO_2 used for EOR and stored in the Aquistore Project.

Boundary Dam Power Station and the ICCS Facility

Overview of BD3 Integrated CCS Process

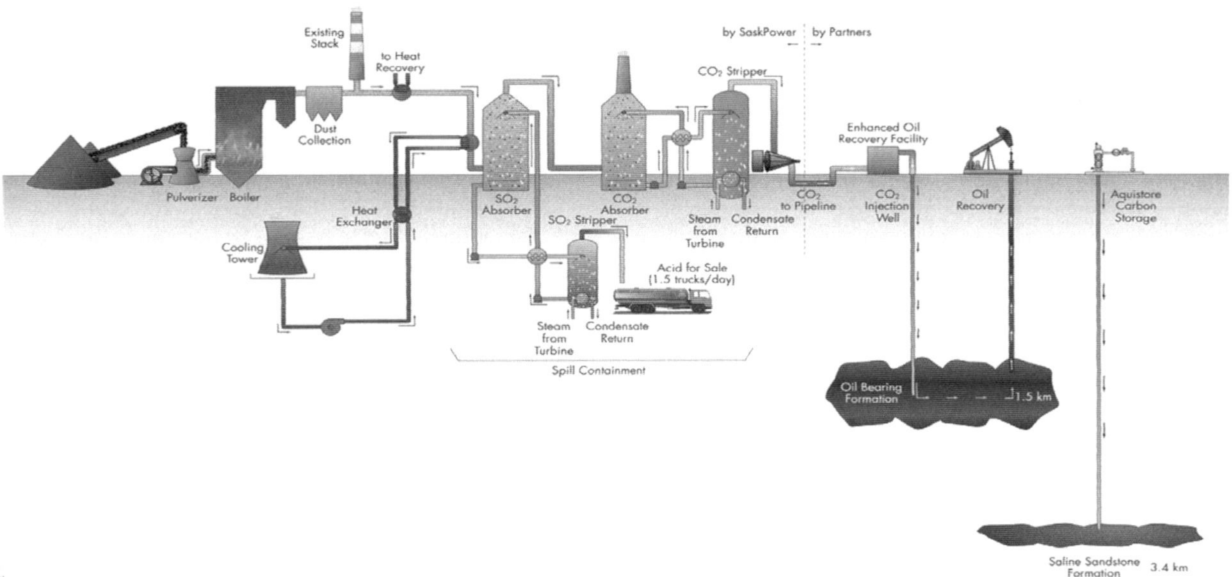

BD3: Successes, Challenges & Valuable Lessons

SUCCESSES

- Pilot testing & 7000:1 scale-up
- Thermal integration

CHALLENGES

- Particulate buildup & amine degradation
- Reduced capacity
- Compressor issues

(Giannaris et. al., 2021)

INTERNATIONAL CCS KNOWLEDGE CENTRE 7

Process Solutions

AMINE DEGRADATION

- Thermal reclaiming unit was reconfigured to increase reclamation capacity

FOULING FROM PARTICULATES

- Spray nozzles added to inlet cooler and booster fan
- Redundant heat exchangers added to enable online maintenance

(Pradoo et. al., 2022)

INTERNATIONAL CCS KNOWLEDGE CENTRE

CO_2 Compressor – Lessons Learned

OUTAGES

- Carbon steel coolers on dry stages, CO2 leaked into cooling water, replaced with stainless steel
- Manufacturing defect in stainless steel cooler[1]
- Motor: loose bolt caused damage – factory quality issue[2]

SPARE PARTS

- Bull gear
- Pinions with Impellers
- Complete set of Seals

(SaskPower Blog, 2017 / Pradoo et. al., 2022)

Recent BD3 Performance

PERFORMANCE HAS STEADILY IMPROVED SINCE STARTING UP AFTER THE COMPRESSOR FAILURE

- Dip in Q2 of 2023 was scheduled maintenance
- 2019 to present: the CCS facility has been able to capture over 90% of the CO_2 from the flue gas it receives
- Carbon intensity of Unit 3 is now around 300 t/GWhr
- Over 6 million tonnes of CO_2 have now been captured

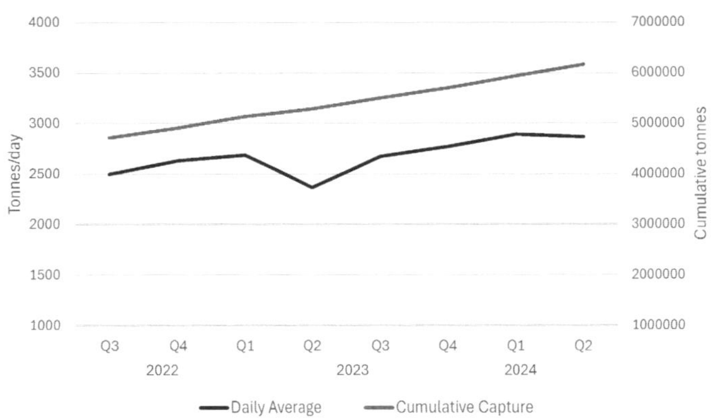

(SaskPower Blog, 2022-2024)

Heidelberg Materials - Edmonton

- Total capture will be 1 million tonnes per year
- Amine post-combustion capture selected
- Currently nearing the end of FEED and piloting on-site
- Offtake partner in place with transport and CO_2 storage

CCS on Cement: Heat Integration

- Heat Integration is not possible to the same extent as a power plant
  - Dedicated steam source is required
  - Supplemental energy from waste heat was studied but was not economical in this case
- Combined heat and power (CHP) improves the business case and the plant dispatch, but increases the cost
- CO_2 capture of **95%** can be achieved with a blended flue gas
- Steam driven compressor decreases parasitic load

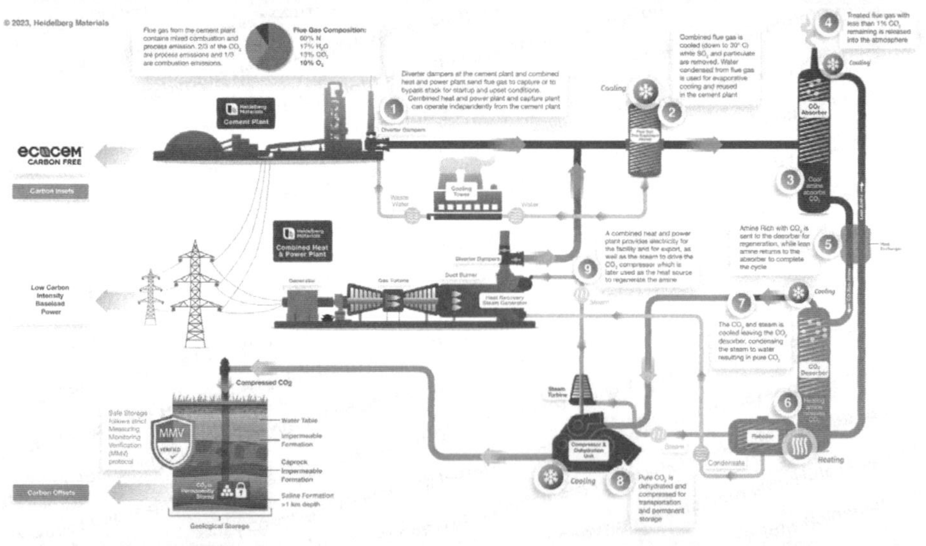

(Tantikhajorngosol et. al., 2024)

Impact of CHP Plant on CO_2 Capture Plant Size

- Adding the flue gas from CHP to the kiln flue gas fed to the CCS facility will change the flue gas flow, CO_2 concentration, and CO_2 quantity, requiring changes to the CCS facility design
- This will result in, among other things, increased steam consumption. This means the sizing of the CHP and CCS plant is an iterative process

Questions?

INTERNATIONAL
CCS KNOWLEDGE
CENTRE

lwestover@ccsknowledge.com
ccsknowledge.com

References

- Giannaris, S., Janowczyk, D., Ruffini, J., Hill, K., Jacobs, B., Bruce, C., Feng, Y., Srisang, W., "SaskPower's Boundary Dam Unit 3 Carbon Capture Facility - The Journey to Achieving Reliability" (January 15, 2021). Proceedings of the 15th Greenhouse Gas Control Technologies Conference 15-18 March 2021, Available at SSRN: https://ssrn.com/abstract=3820191

- Pradoo, P., Jacobs, B., Hill, K., Wilkes, S., Ruffini, J., Srisang, W., Feng, Y., Daverne, D., Nelson, C., "Improving the Operating Availability of the Boundary Dam Unit 3 Carbon Capture Facility" (October 24, 2022). Proceedings of the 16th Greenhouse Gas Control Technologies Conference (GHGT-16) 23-24 Oct 2022, Available at SSRN: https://ssrn.com/abstract=4286503

- SaskPower Blog, 2017 to 2024: https://www.saskpower.com/about-us/our-company/blog

- Tantikhajorngosol, P., Srisang, W., Jacobs, B., Cheruvallath, P., Bruce, C., Anwar, R., "Study of Combined Heat and Power Integrated Carbon Capture and Compression of Heidelberg Materials Edmonton CCUS Project", Mitacs and PTRC Press Conference, Regina, Canada, February 2024.

国家能源集团
CHN ENERGY

The 41st International Pittsburgh Energy and Carbon Management Conference (PECMC-2024)

NATIONAL INSTITUTE OF CLEAN AND-LOW-CARBON ENERGY
北京低碳清洁能源研究院

Application of Composite External Donor Technology in Coal-Based Polypropylene Production

Peiqian Yu, Ph. D

National Institute of Clean-and-Low-Carbon Energy (NICE), CHN Energy (CE), Beijing, China

Oct. 7 -9, 2024, Pittsburgh, PA, USA

Outline

- Introduction to Coal-Based Polypropylene (PP)
- Development of Composite External Donor (ED) Technology at Lab Scale
- Application of ED Technology at Commercial PP Plant
- Conclusions

2

Introduction to Coal-based Polypropylene (PP)

- **Technology Enabling Conversion of Coal to PP in China**
  - **Coal-based PP is an extension** of coal chemical industrial production chain
  - 1st **commercial production** was achieved in 2011 in China, and has experienced **a steady growth**
  - Coal-based polyolefin (PP & PE) is the 2nd **largest product** in the coal chemical industry products , next to liquid oil products

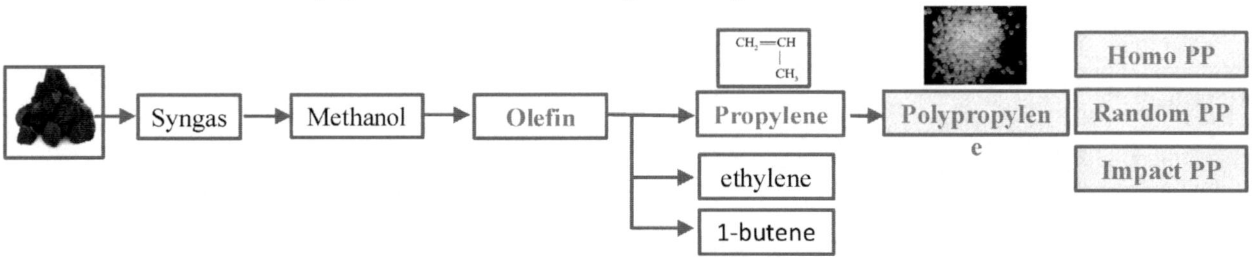

Introduction to Coal-Based Polypropylene (PP)

- **Coal Chemical Industrial Production Chain Extended to Coal-Based PP Production**

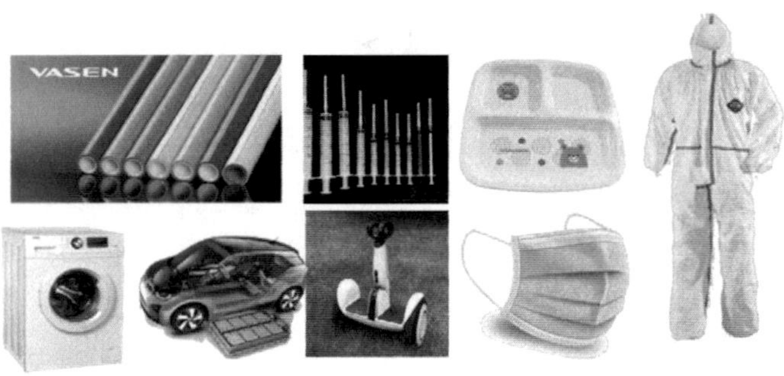

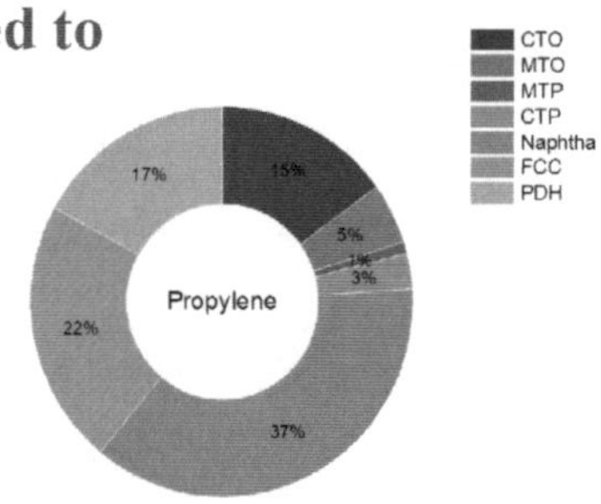

2024 PP Capacity in China

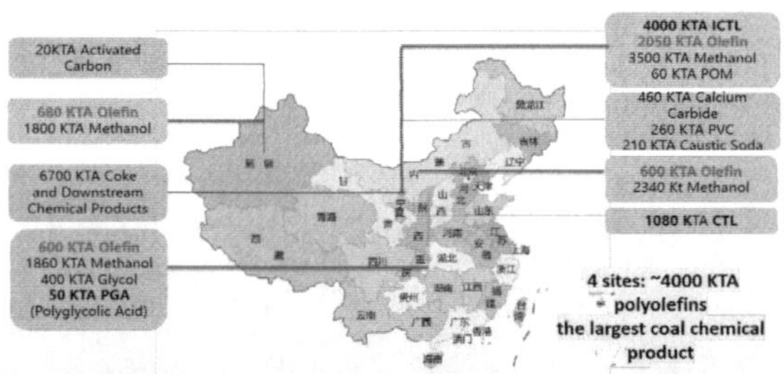

Catalysts and Effects of External Donors (ED)

- **4th Z-N: $TiCl_4$/ID/$MgCl_2$**
- **Alkylaluminum: $Al(Et)_3$**
- **External donors (EDs) are mainly various alkoxy-substituted silane compounds**

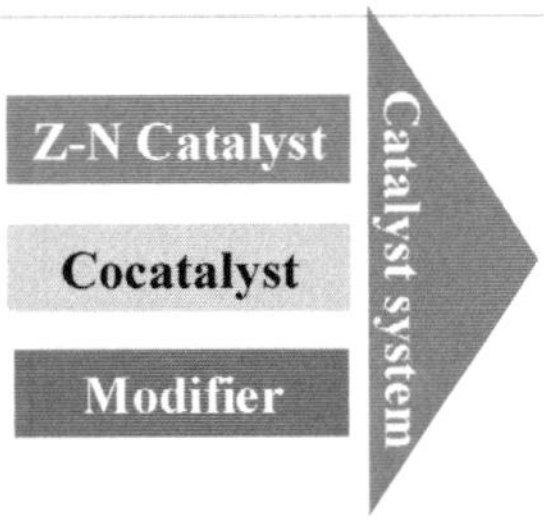

- Catalyst activity
- Hydrogen sensitivity
- Stereo-regularity
- ...

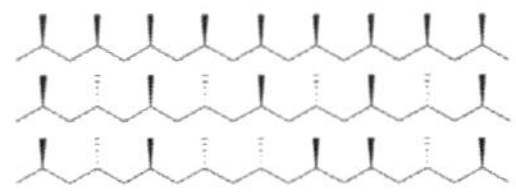

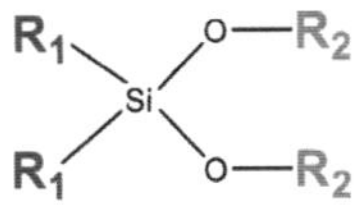

R1= CH_3, n-butyl, isobutyl, Ph

R2 = CH_3, Ethyle, isopropyl

Effect on Stereo-regularity (Isotacticity of PP)

- **R_1(steric hindrance) :**
  - Methyl < n-butyl < isobutyl ≈ phenyl < isopropyl
- **R_2(electron donating ability):**
  - Methyl > Ethyl > isopropyl

5

Develop Composite ED Technology at Lab

- **D/T Pair: dicyclopentyldimethoxysilane (D) & tetraethoxysilane (T)**
- **P/T Pair: diisopropyldimethoxysilane (P) & tetraethoxysilane (T)**

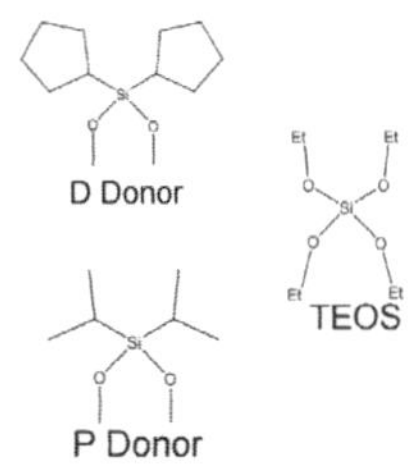

☑ **Polymerization conditions**

- **Propylene homopolymerization**
- **Slurry process**
- **@70±1°C,**
- **@~3.4MPa**
- **1 hour**

☑ **Testing index**

- **Catalyst activity**
- **Melting Flow Rates (MFR)**
- **Xylene Soluble Content (XS%)**
- **Melting temp by DSC (T_m)**

6

Develop D/T Composite ED at Lab: Effect of H_2/C_3

- ✓ **Both catalyst activity and XS% changed non-monotonically with H_2/C_3**
- ✓ **MFR increased dramatically with H_2/C_3**
- ✓ T_m changed slightly
- ☐ **At higher H_2/C_3, catalyst performance changed in a complex manner**

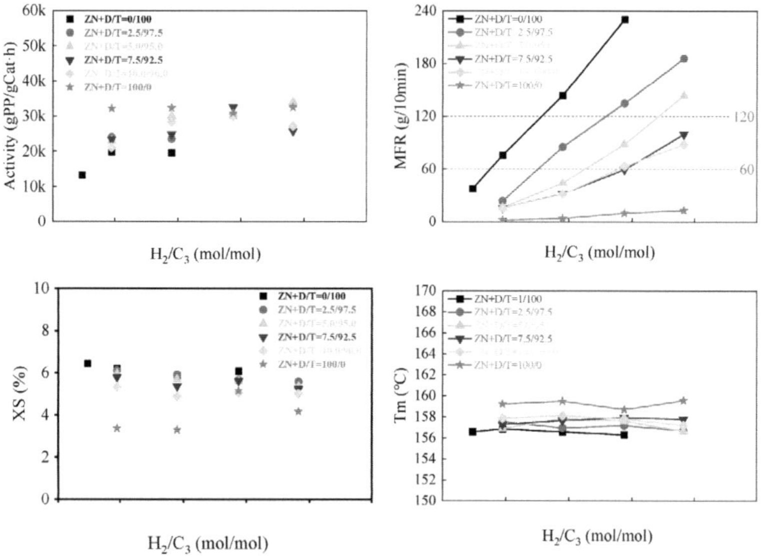

7

Develop D/T composite ED at Lab： D/T Ratio

- ✓ Catalyst activity changed slightly with the increment of D/T
- ✓ **Both MFR and XS% decreased obviously with D/T increasing**
- ✓ **T_m increased with D/T**
- ☐ **Consequently, D/T remarkably affects XS%, MFR and T_m**

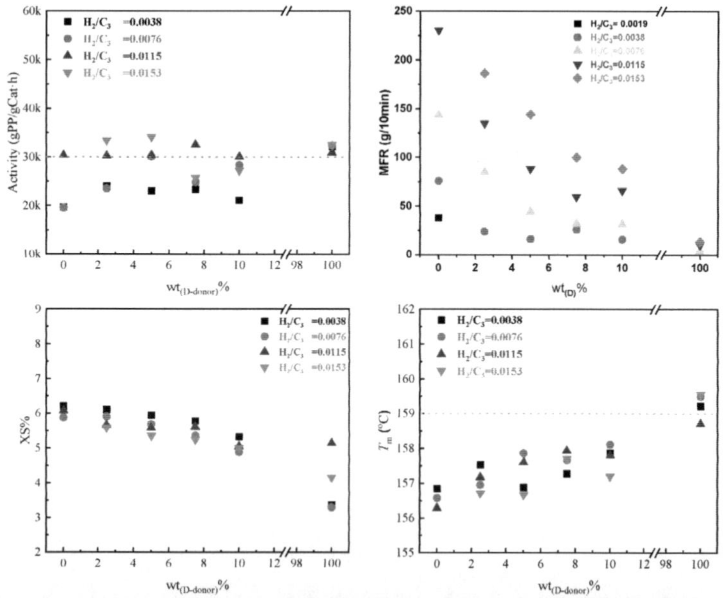

8

Composite P/T ED Tech at Lab: Effect of H_2/C_3

- ✓ **Catalyst activity increased non-monotonically with H_2/C_3**
- ✓ **MFR increased dramatically with increasing H_2/C_3**
- ✓ XS% & T_m changed slightly

□ **H_2/C_3 affects catalyst activity and MFR remarkably, similar trend to D/T**

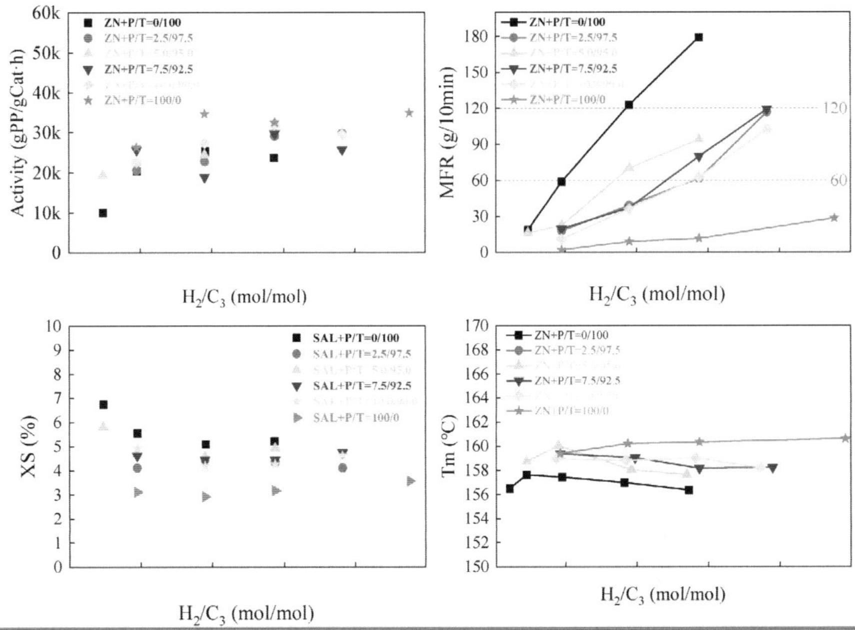

Develop P/T composite ED at Lab: P/T Ratio

- ✓ Catalyst activity changed slightly with increment of P/T ratio
- ✓ **MFR decreased obviously**
- ✓ **XS% dramatically decreased with P/T also**
- ✓ **T_m increased overall**

□ **P/T remarkably affects XS%, MFR and T_m**

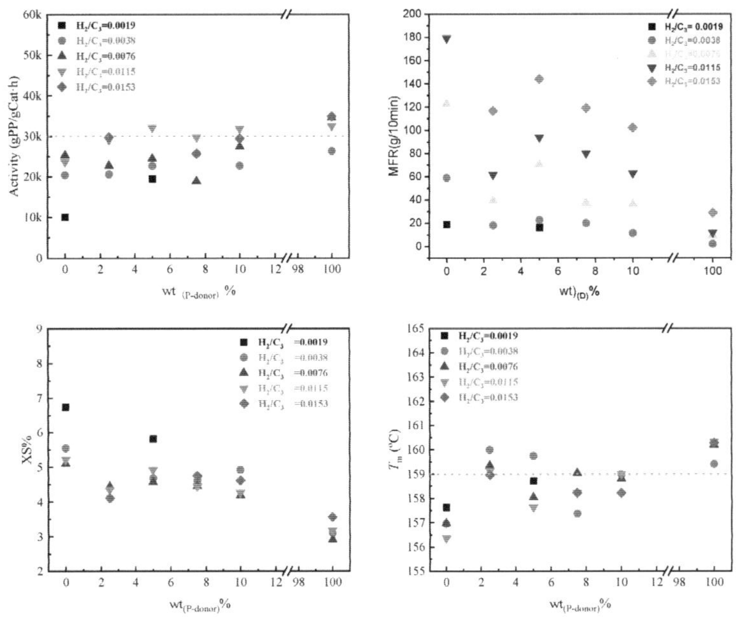

Development of Composite ED Tech at Lab

- **D/T: radar analysis**
- **P/T: radar analysis**

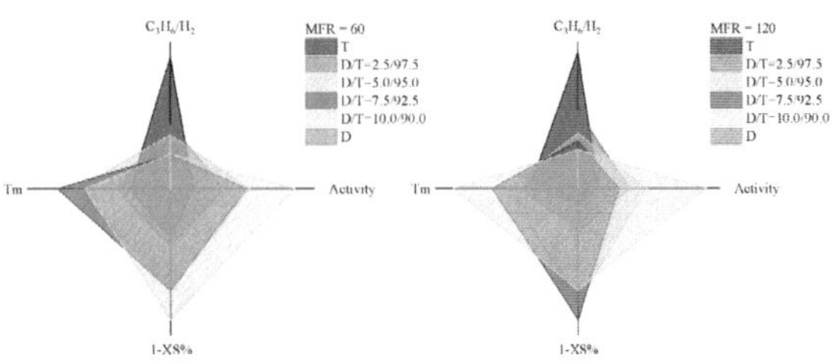

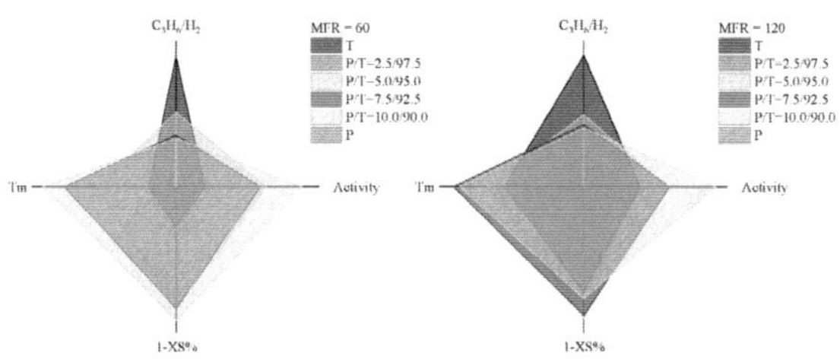

➢ With increasing strong D/P, both activity and stero-regularity increased, however the H_2 sensitivity was going down

➢ **Under higher H_2 condition**, when D/P increased to over 7.5%, **the stero-regularity decreased,** which means the catalyst performance will **drop off**

Modeling P/T Recipes with six-sigma Tools

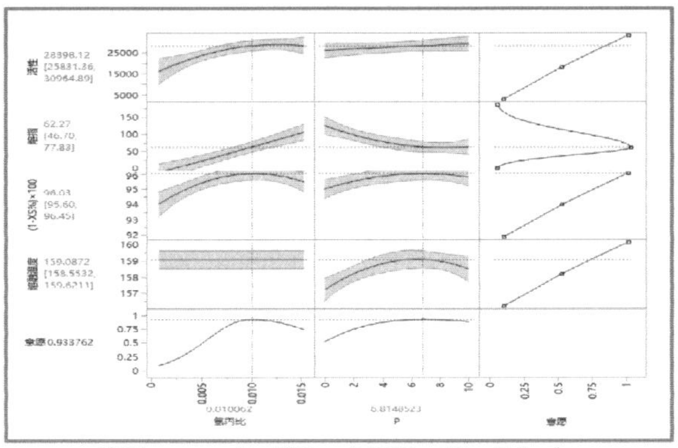

P/T comosite EDs: **MFR=60** g/10min

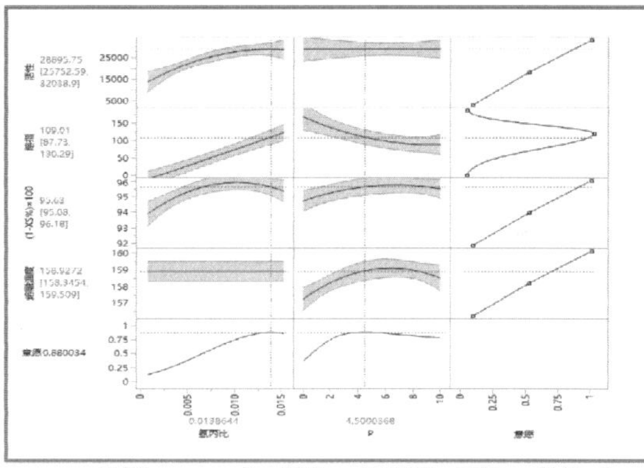

P/T comosite EDs: **MFR=120** g/10min

➢ **P/T** system has a maximum equivalent stereo-regularity

➢ Strong ED, like P-donor acts differently at low or high H_2

➢ **P-donor strongly affects stereo-regularity under low H_2, but not at higher H_2**

Optimization of Composite ED Recipes

Target MFR g/10min	Composite EDs	Ratio(Propietary)	H_2/C_3 mol/mol
60	P/T	★/★	0.010
120	P/T	★/★	0.013
60	D/T	★/★	0.014
120	D/T	★/★	0.015

- **It is necessary to** adjust the ratio of composite ED in real time **slightly based on the target MFR, for meeting the requirements of high performance PP resins**

13

Further Verification with Optimized P/T Recipes at Lab

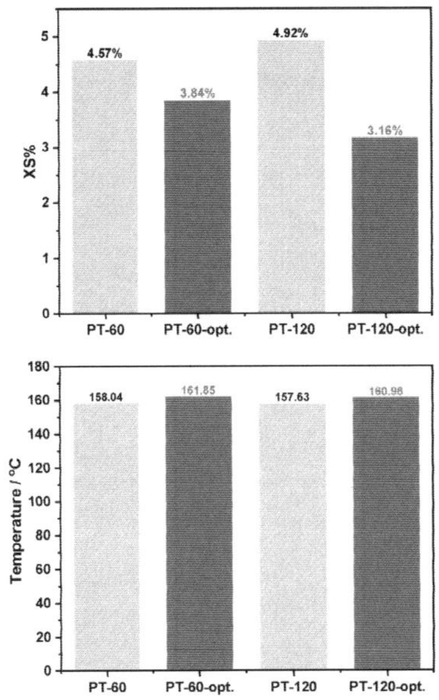

XS% < 4%: **decreased by 0.8%~1.7%**

T_m > 160 °C: **increased by 3.3~3.8 °C**

★ Higher isotacticity and stiffness ↑

14

Succesful Implementation of ED at Commercial PP Plant

- High MFR PP Industrial production at a 4.5 million tons/yr with optimized recipe

No.	P/T mol/mol	Notched Izod impact strength KJ/m^2	Tensile strength MPa	breaking elongation %	Flexural modulus MPa
S1	☆/☆	14.2	24.3	112.2	1494
S2	★/★	12.1	24.9	64.8	1591 ↑

★ ~100MPa increased in FM

□ Much more suitable for producing thin-wall injection mold PP products

15

Conclusions

(1) Unique composite ED technology was developed specifically for coal-based PP process.

(2) It was successfully implemented at our coal-based PP plant, where PP products had an improved flexible modulus by ~ 100 MPa.

(3) Path forward: to tailor composite ED to different PP products by coal-based PP production.

16

Acknowledgement

Welcome to NICE, CE, China!

(Tel) +86-10-57595665
(Mobile) +86 10-139-1174-7645
(Email) peiqian.yu@chnenergy.com.cn

Forty-First Annual International Pittsburgh Energy and Carbon Management Conference (PECMC)

Efficient separation of monocyclic aromatics from direct coal liquefaction

Xiaolong Shi, Yuncai Song, Jie Feng, Wenying Li*

Taiyuan University of Technology, China

PECMC, Pittsburgh, America, October 07-09, 2024

1 Background

Separation of DCL

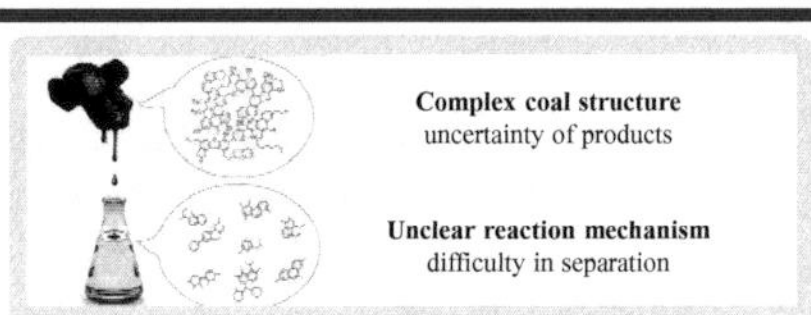

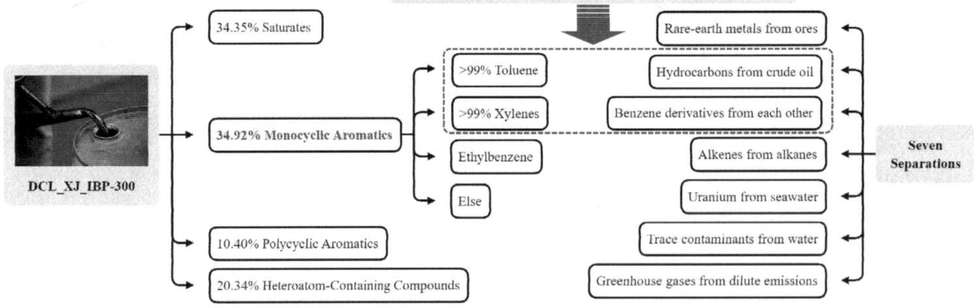

- Direct coal liquefaction(DCL) has more aromatics, and monocyclic aromatics make up the largest proportion of them.

[1] Sholl, D., Lively, R. Nature. 2016, 532: 435–437. [2] Isao Mochida. Chemical Reviews

1 Background

Composition characteristics:

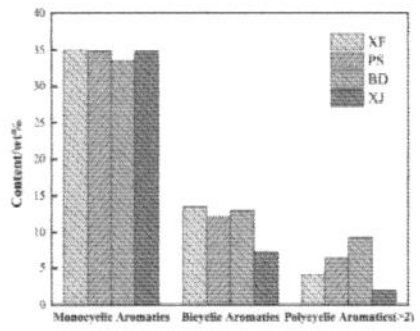

Fig.2 Percentage of aromatics in different DCL

Energy consumption:

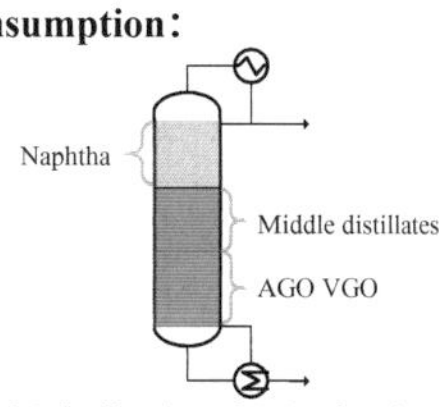

Fig.2 Distribution of DCL fractions

Market demand:

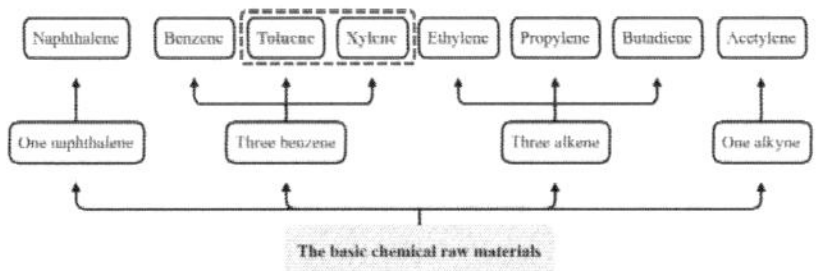

Fig.2 The eight basic chemical raw materials

High-value utilization:

Tab.2 Comparison of prices of different products

	Product	Economic benefits /(10^4yuan/t)	Times	Manufacturers
	DCL	0.35	/	/
High Purity	Toluene (99%)	21.98	60	Aladdin
	Ethylene (99%)	10.89	30	Aladdin
	Xylene (99%)	13.43	40	Aladdin

2 Idea & Method

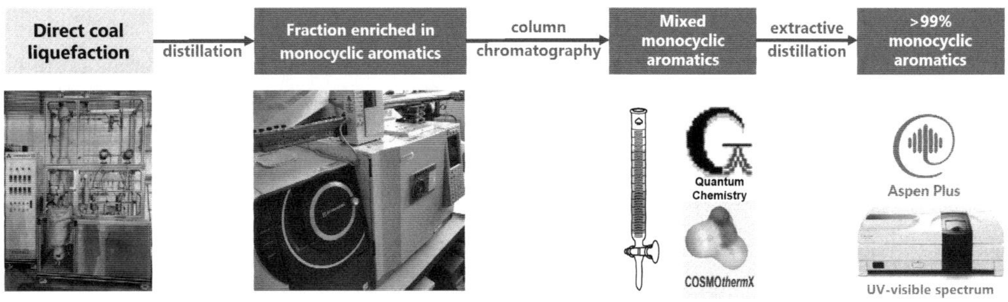

Fig.1. Separation of mono-aromatics from direct coal liquefaction (DCL)

3.1 Distillation

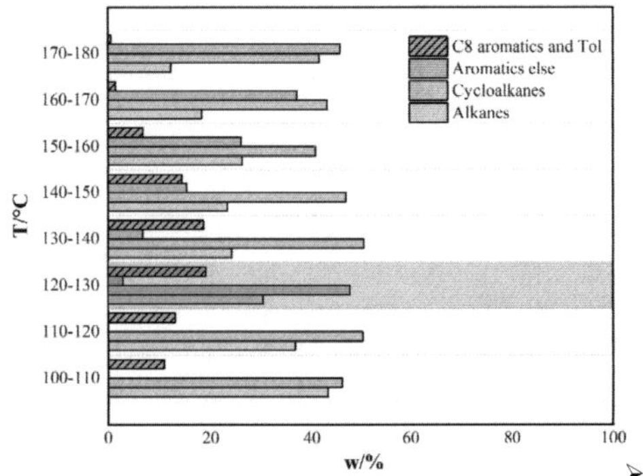

Fig.2. Composition analysis for the fractions

T/°C	Toluene	Ethyl-benzene	Xylenes	Phenol	Else
170-180	N/A	N/A	√	√	√
160-170	N/A	N/A	√	√	√
150-160	N/A	√	√	√	√
140-150	N/A	√	√	√	√
130-140	√	√	√	√	√
120-130	√	√	√	√	N/A
110-120	√	√	√	√	N/A
100-110	√	N/A	N/A	N/A	N/A
IBP-100	N/A	N/A	N/A	N/A	N/A

- The types of aromatics in this fraction are relatively simple, which only includes Toluene, Ethylbenzene, Xylene and Phenol.
- C8 aromatics and Toluene are the most abundant in 120-130 °C

3.2 Column chromatography

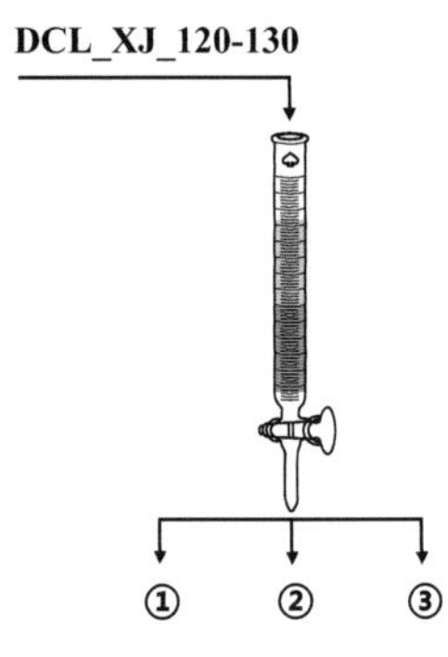

Fig.1 Schematic diagram of separation device

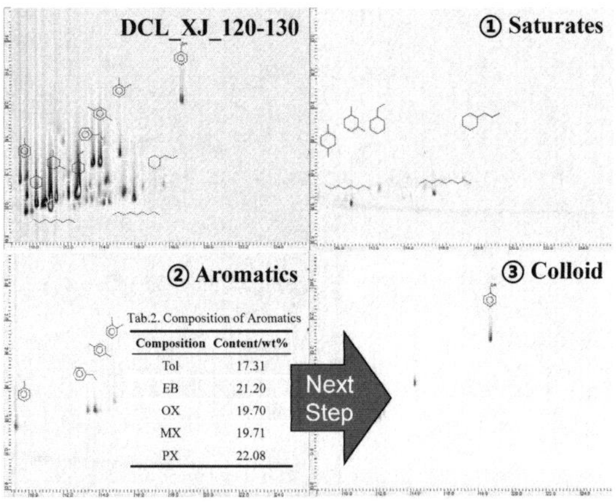

Fig.2 Analysis of separation products by GC×GC-MS

- Aromatics are separated from this fraction (120-130 °C)

3.3 Extractive distillation

Tab.1. Boiling points of aromatics

Composition	Content/wt%	B.P./°C
Toluene	17.31	110.6
EB	21.20	136.1
PX	22.08	138.4
MX	19.71	139.3
OX	19.70	144.4

- Toluene can be separated by distillation;
- It is hard to separate C8 aromatics by distillation;

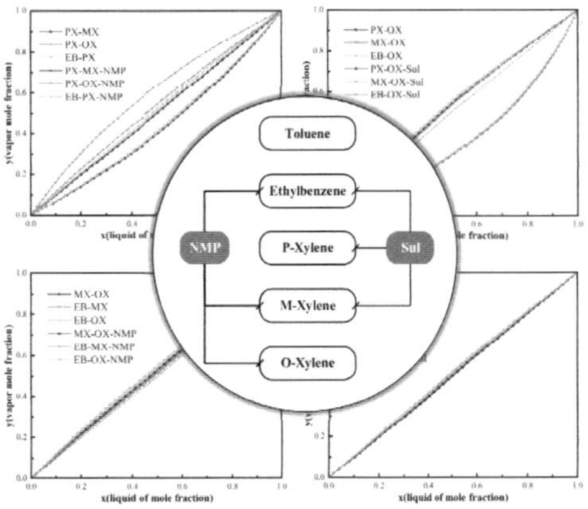

Fig.1.Effect of NMP/Sul on the Vapor-Liquid Equilibrium (VLE) of C8 aromatics

- NMP/Sulfolane(Sul) can increase the relative volatility between para-xylene/ortho-xylene and other components, So we choose NMP and Sul as solvents for extractive distillation to separate p-xylene and o-xylene.

3.4 Final Separation Process

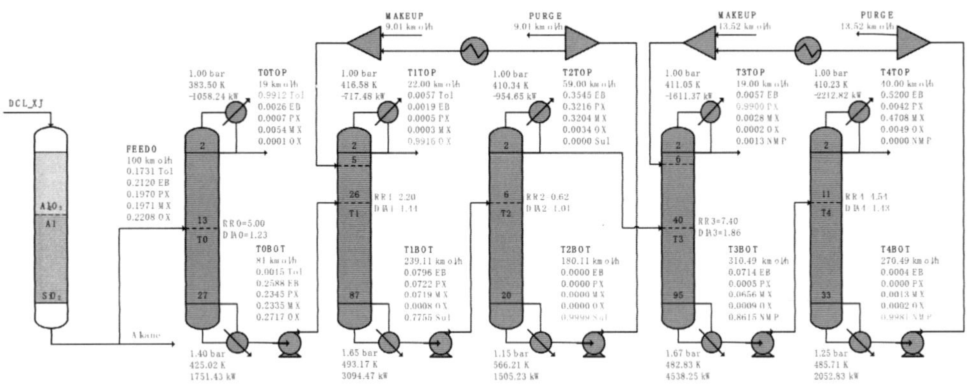

Fig.1. Separation process of monocyclic aromatics in DCL

A1: Adsorption separation, **T0**: Distillation column, **T1/T3**: Extractive distillation, **T2/T4**: Solvent recovery

3.4 Final Separation Process

Tab.1. Composition of process stream

Compound	FEED0	T0-TOP	T1-TOP	T2-BOT	T3-TOP	T4-BOT
TOLUE-01	16.91%	99.12%	0.57%	0.00%	0.00%	0.00%
ETHYL-01	21.54%	0.26%	0.19%	0.00%	0.57%	0.04%
P-XYL-01	19.49%	0.07%	0.05%	0.00%	99.00%	0.00%
M-XYL-01	19.49%	0.54%	0.03%	0.00%	0.28%	0.14%
O-XYL-01	22.57%	0.01%	99.16%	0.00%	0.02%	0.02%
N-MET-01	0.00%	0.00%	0.00%	0.00%	0.13%	99.80%
SULFO-01	0.00%	0.00%	0.00%	99.99%	0.00%	0.00%

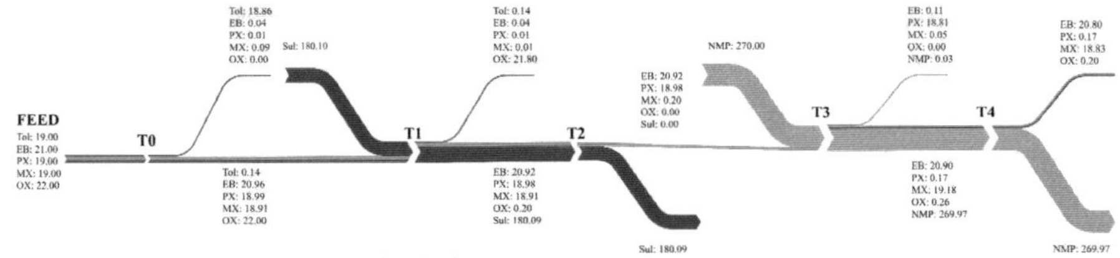

Fig.1. Flow rates of the process stream

- In the end, we get 3 monocyclic aromatics (Toluene, o-xylene, and p-xylene) ,and their purity and yield are ≥ 99%.

3.5 Separation mechanism: Column Chromatography

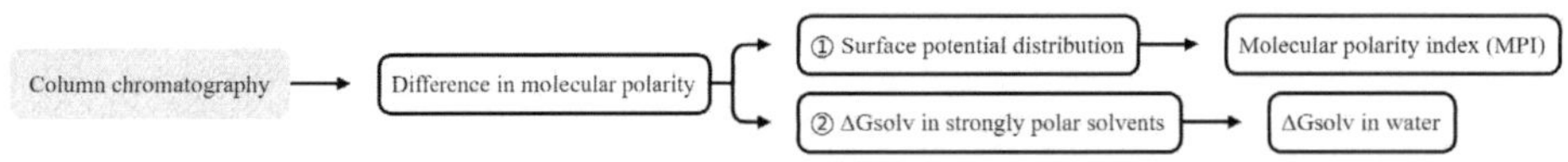

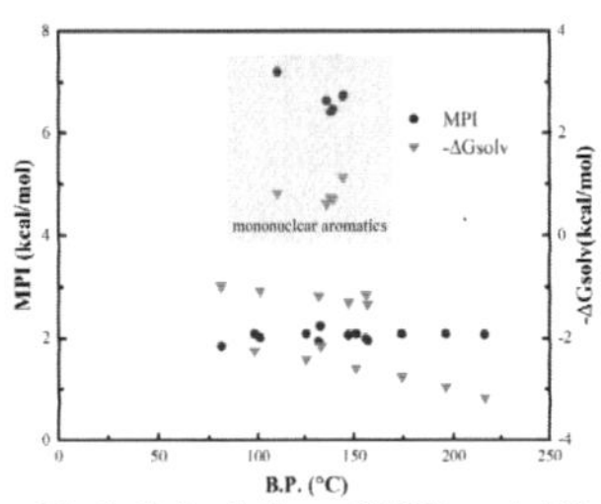

MPI (molecular polarity index):

$$MPI = (1/A)\iint_S |V(r)| dS$$

ΔG_{solv}^{i} (Gibbs free energy of solvation) :

$$\Delta G_{Solv}^{i} = (\mu_{S,\infty}^{i} - \mu_{gas}^{i})$$

Fig.1 Calculation of MPI and ΔG_{solv} for each component in DCL_XJ_120-130

- Aromatics have higher polarity than saturates → The difference in polarity is an effective driving force to separate aromatics.

3 Result & Discussion

3.5 Separation mechanism: Extractive Distillation

C8 aromatics

The effect of the solvent in Extractive Distillation
- **① affect the interaction between components** → **UV-vis spectrum**
- ② dilution (non-ideal system)

Solvent selectivity equation:

$$\ln\left(\frac{\alpha_S}{\alpha}\right) = x_S\left[A_{1S}' - A_{2S}' - A_{12}'(1-2x_1')\right]$$

① Margules equation parameter:

A_{12}' reflects the interaction between the components of the original solution,

A_{1S}', A_{2S}' reflects the interaction between the solvent and the components of the original solution,

② x_S is the molar fraction of solvent,

➢ **The effect of NMP/Sul on C8 aromatics can be explored by UV-visible spectrum.**

3 Result & Discussion

3.5 Separation mechanism: Extractive Distillation

π→π* (B bond, 230-270nm)

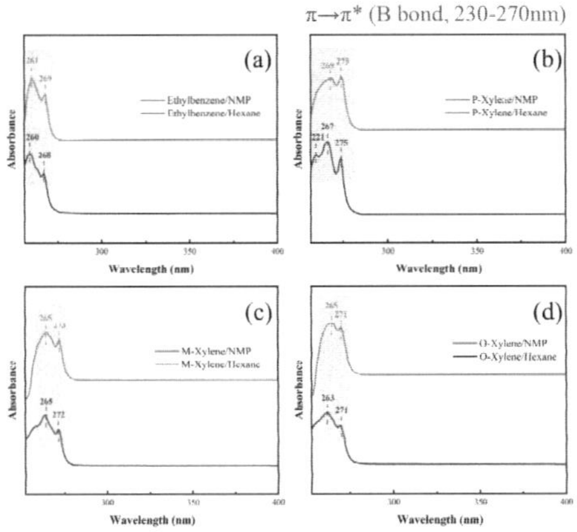

Fig.1 UV-Vis spectra: NMP-EB (a), NMP-PX (b), NMP-MX(c) and NMP-OX(d)

Tab.1 Statistics of the highest absorption peaks

	Hexane	NMP	$\Delta\lambda_{max}$/nm
Ethylbenzene	260	261	+1
P-Xylene	267	275	+9
M-Xylene	265	265	+0
O-Xylene	263	265	+2

➢ The peaks of p-xylene is significantly red-shifted by NMP,

3.5 Separation mechanism: Extractive Distillation

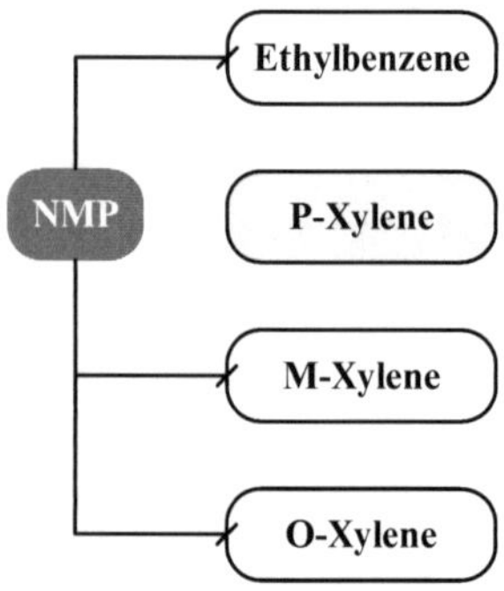

Fig.2 Schematic diagram of interactions

Tab.1 Statistics of the highest absorption peaks

	Hexane	NMP	$\Delta\lambda_{max}$/nm
Ethylbenzene	260	261	+1
P-Xylene	267	275	+9
M-Xylene	265	265	+0
O-Xylene	263	265	+2

- The peak of P-Xylene is significantly red-shifted by NMP,
- The interaction between NMP and P-Xylene is significantly weaker than that between NMP and other components.
- The addition of NMP can promote the separation of p-xylene

3.5 Separation mechanism: Extractive Distillation

n→π* (R bond, >280nm)

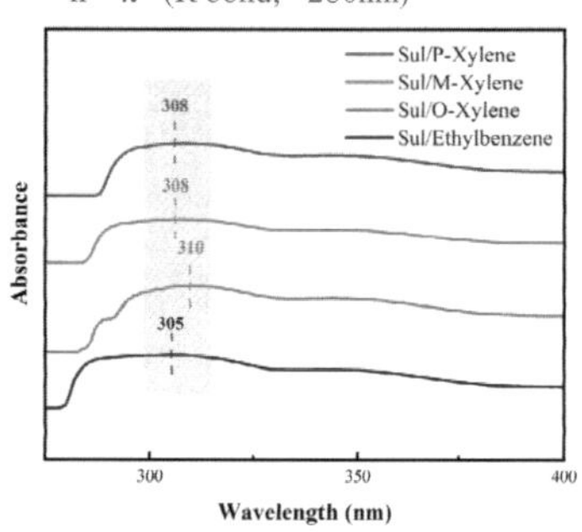

Fig.1 UV-Vis spectra of Sul-C8 aromatics

- The peak of Sul is significantly red-shifted by O-Xylene.

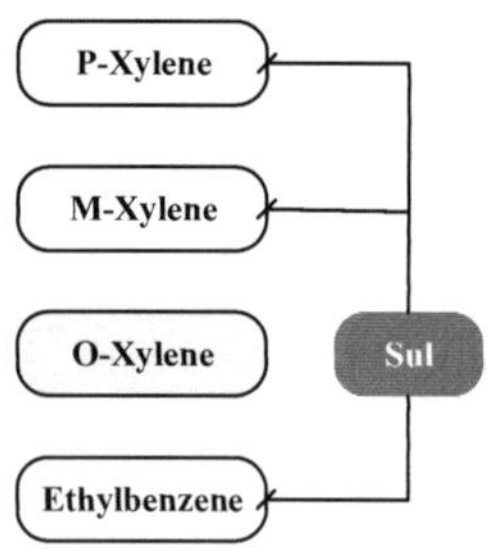

Fig.2 Schematic diagram of interactions

- The interaction between Sul and O-Xylene is significantly weaker than that between Sul and other components.
- The addition of Sul can promote the separation of O-Xylene.

4 Conclusions

- A separation strategy of "**Distillation-Column Chromatography-Extractive Distillation**" was proposed, which can separate **3** monocyclic aromatics from DCL with purity and yield **≥ 99%**;

① C8 aromatics and Toluene are the most abundant in **120-130** °C.

② The mixed aromatics in 120-130 °C could be separated by **Column Chromatography**.

③ O-Xylene and P-Xylene could be separated by **Extractive Distillation** with the addition of Sul and NMP;

- Separation Mechanism:

① **MPI and ΔG_{solv}:** the polarity difference is the effective driving force to separate the mixed aromatics.

② **UV-visible spectra:** the interaction of NMP - P-Xylene < NMP - Others, Sul - O-Xylene < Sul - Others, which is an effective driving force to separate the C8 aromatics;

5 Acknowledgements

Financial Supports

The authors gratefully acknowledge the financial support from the National Key Research and Development Program of China (2023YFB4103404-03).

Thanks for your attention!

State Key Laboratory of Clean and Efficient Coal Utilization

Taiyuan University of Technology, P.R. China

Yuncai Song: songyuncai@tyut.edu.cn

Xiaolong Shi: shixiaolong1056@link.tyut.edu.cn

17

EOR Driven by Wafer Scale Engines

NATIONAL ENERGY TECHNOLOGY LABORATORY

Vyacheslav Romanov
Physical Scientist, NETL

Disclaimer

NATIONAL ENERGY TECHNOLOGY LABORATORY

- This report was prepared as an account of work sponsored by an agency of the United States Government. Neither the United States Government nor any agency thereof, nor any of their employees, makes any warranty, express or implied, or assumes any legal liability or responsibility for the accuracy, completeness, or usefulness of any information, apparatus, product, or process disclosed, or represents that its use would not infringe privately owned rights. Reference therein to any specific commercial product, process, or service by trade name, trademark, manufacturer, or otherwise does not necessarily constitute or imply its endorsement, recommendation, or favoring by the United States Government or any agency thereof. The views and opinions of authors expressed therein do not necessarily state or reflect those of the United States Government or any agency thereof.

UNet Performance with Wafer Scale Engine

Presentation Outline

1. Background Overview
   a) Problem statement
   b) State-of-the-art
2. Workflow
   a) Model template
   b) Prototype
3. Case Study: CO_2-EOR
   a) Data wrangling
   b) Compute metrics
4. Summary

Image source: https://www.zdnet.com/

Massive acceleration of subsurface simulations

Complex Oil/Gas Field Operations Require Fast Decision Support

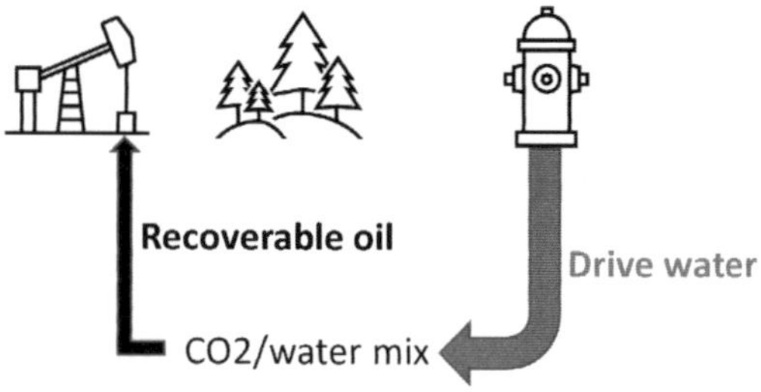

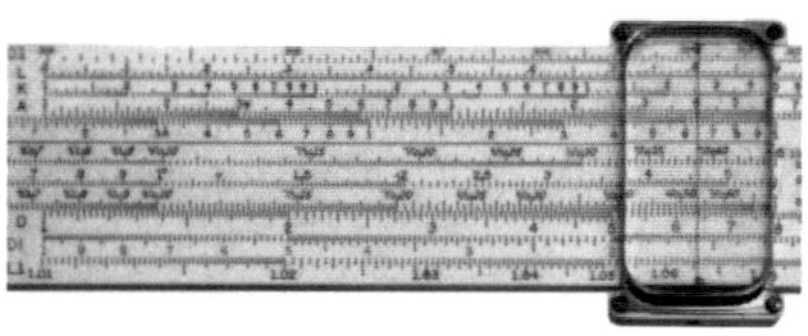

Applications
- Complex problems
- Need for fast compute

Computations
- Hardware limitations
- Long compute time

Hardware acceleration specifications

NATIONAL ENERGY TECHNOLOGY LABORATORY

Cerebras CS-2 (7 nm) vs NVIDIA GPU (4 nm)

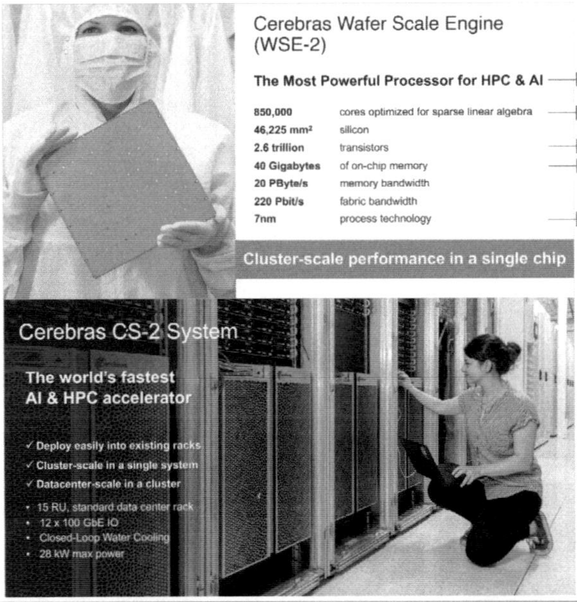

SPECIFICATIONS	Baseline			H200
	H100 SXM	H100 PCIe		
FP64	30 TFLOPS	24 TFLOPS		
FP64 Tensor Core	60 TFLOPS	48 TFLOPS		
FP32	60 TFLOPS	48 TFLOPS		
TF32 Tensor Core	1,000 TFLOPS*	800 TFLOPS*		
BFLOAT16 Tensor Core	2,000 TFLOPS*	1,600 TFLOPS*		
FP16 Tensor Core	2,000 TFLOPS*	1,600 TFLOPS*		
FP8 Tensor Core	4,000 TFLOPS*	3,200 TFLOPS*		
INT8 Tensor Core	4,000 TOPS*	3,200 TOPS*		
GPU memory	80GB	80GB	→	141 GB / 6-stack
GPU memory bandwidth	3TB/s	2TB/s	→	4.8 TB/s
Decoders	7 NVDEC 7 JPEG	7 NVDEC 7 JPEG		
Max thermal design power (TDP)	700W	350W		
Multi-Instance GPUs	Up to 7 MIGS @ 10GB each			
Form factor	SXM	PCIe dual-slot air-cooled		
Interconnect	NVLink: 900GB/s PCIe Gen5: 128GB/s	NVLink: 600GB/s PCIe Gen5: 128GB/s		
Server options	NVIDIA HGX™ H100 partner and NVIDIA-Certified Systems™ with 4 or 8 GPUs NVIDIA DGX™ H100 with 8 GPUs	Partner and NVIDIA-Certified Systems with 1–8 GPUs		

* Shown with sparsity. Specifications 1/2 lower without sparsity.

Model Development Workflow

NATIONAL ENERGY TECHNOLOGY LABORATORY

Computational Environment

Neocortex/Cerebras integration

- Federated w/Bridges-2
  - - Ocean file system
- Ethernet links (data flow)
- Scaling across multiple WSEs

HPE Superdome Flex/CPU only

- Login node (32 CPUs)
- SDF nodes (interactive sessions)
  - Resource allocation
  - Singularity (Python 3.7)

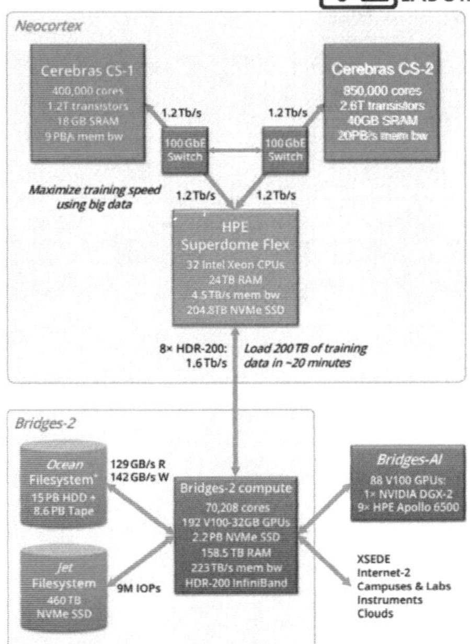

Model Development Workflow

NATIONAL ENERGY TECHNOLOGY LABORATORY

Cerebras CSoft platform and Graph Compiler stages

- Cerebras Estimator (TensorFlow)
- Intermediate Representation
- High-performance kernels
- Linear Algebra operators
- Auto-generated code
- Kernel graph mapping to fabric
- Data routing (streaming)
- Executable
- State machines:
  - training and auto-correction
  - redundancy (for defects)

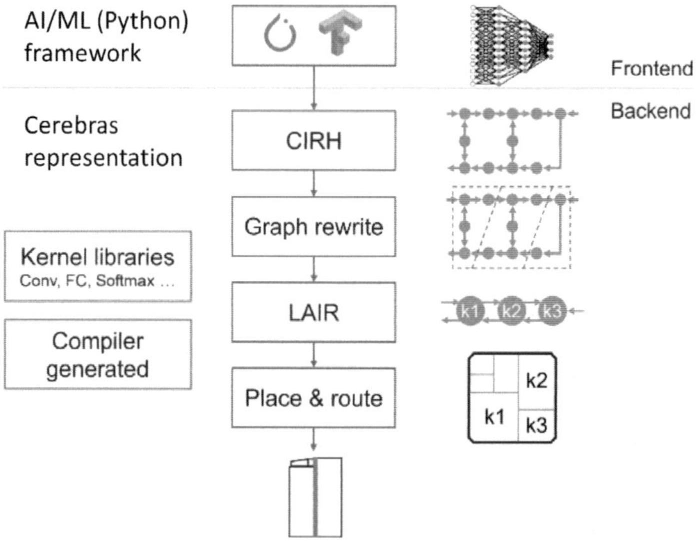

Model Development Workflow

NATIONAL ENERGY TECHNOLOGY LABORATORY

UNet Model (TensorFlow)

PSC/Cerebras template (experimental)

- UNet layers matched to kernels
- Dataset class
- Classification/segmentation
  - BCE loss
  - Dice/mIoU, multi-level PCA

Prototype features

- Tabular input to image
- Loss layer (w/base policies)
- Regression (loss, metrics)

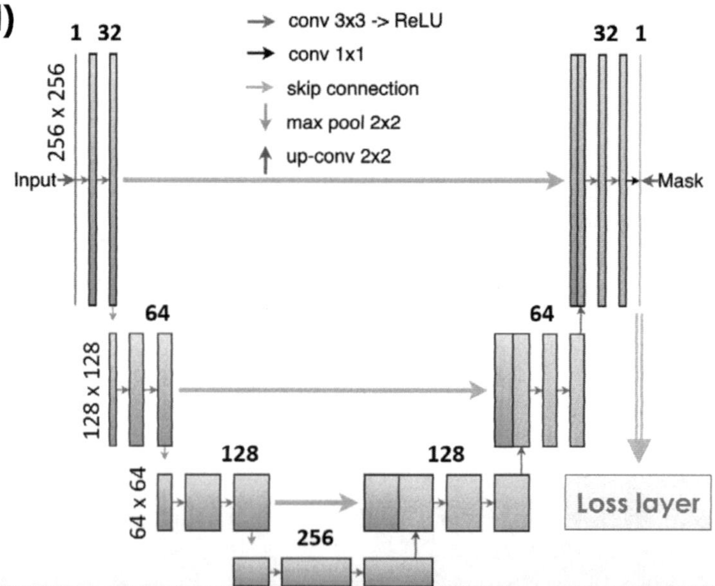

Model Development Workflow

Model training and evaluation

Training (Cerebras WSE)

- Generating spatio-temporal images
- Cerebras compilation
- WSE fabric programming
- Data streaming (pipeline)
- Optimization steps (Adam)

Evaluation (Superdome Flex CPU)

- Interactive session (SDF node)
- Singularity container (Python)
- Inference (test sample)

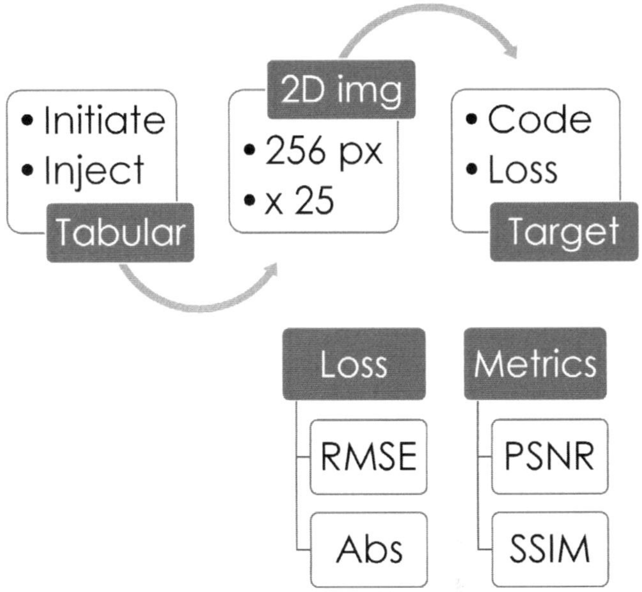

Case Study

CO_2-Enhanced Oil Recovery
AI/ML model training

Case Study

SACROC model (Permian Basin: western TX + southeastern NM)

- Input data 2D mapping/reshaping

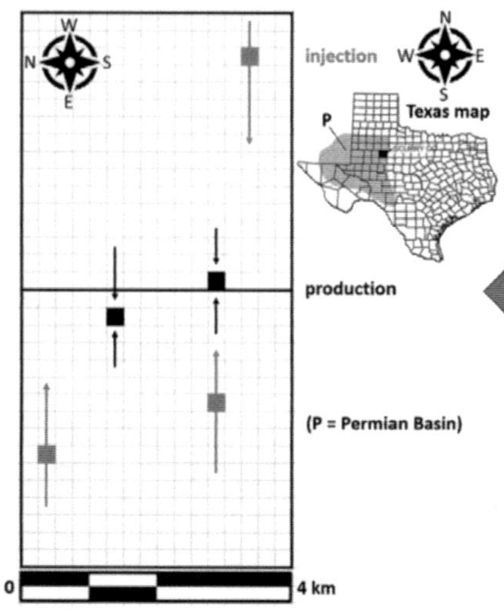

Spatial dim: 16 x 16 to 256 (25 layers)

Spatial dim: 16 x 16 to 256 (25 layers)

Time steps: 922 to 4 x 256 (each year)

- Prior reservoir simulation using CMG GEM simulator produced high fidelity **target data (coarse)**

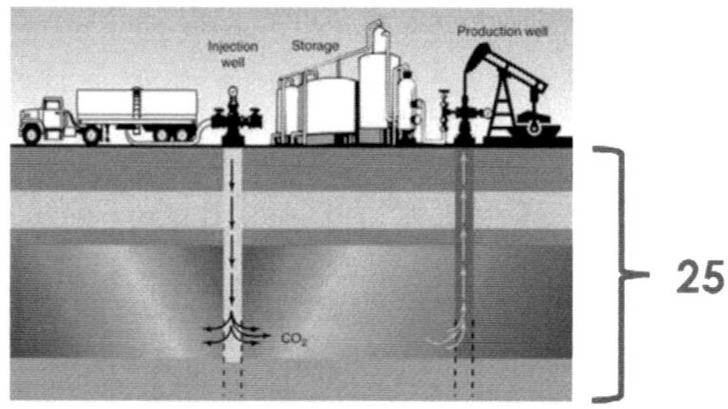

- **Pore pressure (81 + 9 scenarios)**
- **Geologic realizations (A and B)**

Case Study

INCREMENTAL MODEL TRAINING: WSE (train) and CPU (test) metrics

Train Steps	Depth Layers	Inject	Loss Scale	Compute time (sec)
0 – 1,000	0 and 4	with & w/o	x 2^9	11.615
1,000 – 1,250	0 – 7	Yes	Dynamic	7.984
1,250 – 4,000	0 – 7	Yes	x 2^9	39.371
4,000 – 10,000	0 – 15	Yes	x 2^9	86.255

Test Steps	Depth Layers	Inject	CPU Cores	Inference time (sec/step)
1,000	0 – 4	Yes	14 x 28	217.8
1,250	0 – 4	Yes	14 x 28	
4,000	0 – 4	Yes	14 x 28	
10,000	0 – 4	Yes	14 x 28	

- CS-2 (7-nm WSE) number of cores = 850,000
- CS-1 (16-nm WSE) not mapping the AI model

Other relevant WSE training metrics

- CS-2 PEs % active = 60%
- CS-2 Compute utilization = 30%
- CS-2 Fabric Programming time = 149.5 s
- CS-2 Data streaming: 8446.6 samples/s

Summary

NATIONAL ENERGY TECHNOLOGY LABORATORY

- The compute architecture for AI modeling of subsurface was developed based on the experimental **UNet** template available from Cerebras Model Zoo.
- The Model was converted from classification to regression tasks by incorporating the base-layer policies into **Loss Layer**.
- The AI model was successfully mapped to CS-2 WSE and executed in **pipeline mode**.
- **Mixed precision and supervised dynamic loss scaling** were used to overcome the half-precision challenges with convolutional layers.

Acknowledgments

NATIONAL ENERGY TECHNOLOGY LABORATORY

- **Cerebras Systems and Neocortex teams** at the **Pittsburgh Supercomputing Center** provided vital support during the early stages of setting up the computational environment under the **NSF cooperative agreement**.

- **NETL research contractor support** is recognized for assistance with the energy data exchange.

2024 International Pittsburgh Coal Conference
October 7 - 9, 2024
Abstract Submission
PROGRAM TOPIC:
CLEAN DEMONSTRATION AND COMMERCIAL PROJECTS
Exact Title of Paper:

ROCK WOOL SPUN FROM VITRIFIED COAL FLY ASH SLAG IS VERY PROFITABLE

Howard McClintic, Director of REEs & Recycling, Concurrent Technologies Corporation, Washington, DC, U.S.A.
McClintH@ctc.com, (202) 689-4586

ABSTRACT

Concurrent Technologies Corporation (CTC) is headquartered in western Pennsylvania (PA); CTC has a high temperature electric arc furnace (EAF) technology that is very versatile and has many applications. This technology reaches 3500°C (degrees centigrade equals 6332°F) or higher. When exposed to high temperatures, all materials are gasified, reformed into a simpler, more stable material. The organic component (when gasified) becomes a clean syngas, the metals become alloys, and the inorganic material becomes slag, which is a coarse, sand-like material and contains oxides of calcium, silicon, iron, and aluminum, with lesser amounts of manganese, phosphorus, and other materials. Slag is used as an input to produce mineral/rock wool insulation for homes as well as car dashboards; hydroponic blocks are another rock wool product, made from slag. The cost of producing slag is $18/per ton.

There are two manufacturers of rock wool producing equipment – both are in the EU (European Union). Our western PA Partner, Robindale Energy, welcomes the construction and operation of the rock wool manufacturing equipment at one of the three western PA sites that it supplies with fuel, with initial annual tonnage around 100,000 tons. The gross profitability is stated in the two given price points of $30/ton for selling slag for producing cement and $250/ton for manufacturing rock wool, from an identified buyer, growing the (rock wool) business in New England and eastern Canada.

The power plants that Robindale Energy serves deliver electricity to 650,000 homes and also generates alkaline ash that is certified by the Pennsylvania Department of Environmental Protection for beneficial uses. Historically and prospectively, they combust reclaimed coal refuse piles. Coal refuse/gob (garbage of bituminous) is a byproduct waste stream created by over a century old legacy of coal mining and is a water and air pollution nightmare in the Appalachian Region. When snowed and/or rained upon, the acidic drainage water/effluent drains to a creek, stream, or lake, turning that body of water to the color of orange, because of its high iron content. This acid mine drainage (AMD) is reduced with every ton of coal refuse removed. Many of the coal refuse piles also experience spontaneous combustion caused by the exothermic oxidation of sulfur in presence of carbon and oxygen in the materials which then releases carbon and sulfur dioxides, methane, hydrogen sulfide and many other uncontrolled pollutants into the atmosphere.

The PA legislature as well as the PA Department of Environmental Protection (DEP) well know and recognize this major on-going benefit provided by Robindale Energy: over the past twenty years, the company has removed more than 60,000,000 tons of refuse pile coal from the Pennsylvania landscape and further reclaimed the sites' contour, stability, and productivity. This unique supply chain (of reclaimed coal refuse) enables Seward Generation, Colver Green Energy, and Ebensburg Power Company (https://robindale.energy/companies/) to operate as three of the lowest cost solid fuel power plants in the PJM (Pennsylvania/New Jersey/Maryland) Region.

EXECUTIVE SUMMARY

There are two traditional sources of slag: the steel and solid fuel power generating industries. Nationally, because of environmental laws, neither industry will likely expand; the effect of the full implementation of the Paris Climate Agreement is another longer-term inhibiting factor. However, it is unlikely that the Robindale Energy situation will change because of the positive environmental benefits provided as well as the millions of tons of g that they will continue to produce.

CTC is building from scratch a demonstration-sized furnace to display the many capabilities of this gasification technology: processing MSW (municipal solid waste); extracting rare earth elements (REEs) from Pennsylvania (PA) electricity generating utility's ash; converting rare earth oxides (REOs) into REEs from PA (Penn State) and other sources; and, can **vitrify coal refuse to produce** slag at 1650°C whose most valuable end-product is rock wool.

Schemata of CTC's Gasification Technology

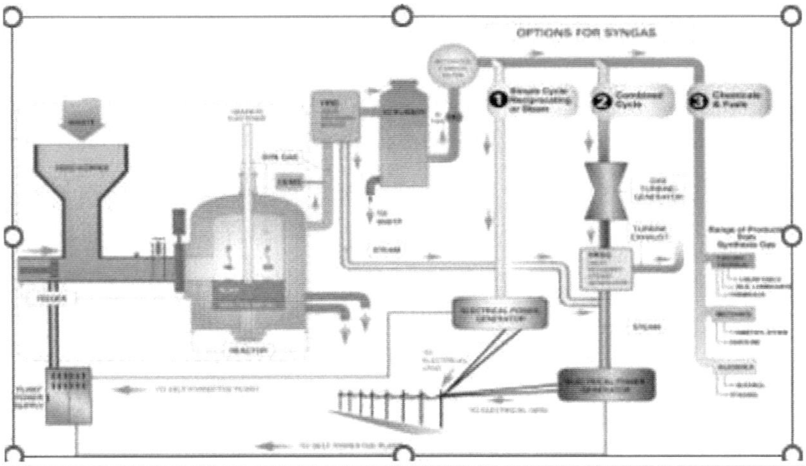

CTC's demonstration high temperature, electric arc furnace (EAF) technology will focus on exceeding the capability of the Chinese to extract REEs (praseodymium, neodymium, samarium) to produce permanent magnets (NdFeB and SmCo). The Chinese EAF technology, provided by Hitachi Metals, has a maximum temperature of 1900°C;; CTC's unit will reach 3500°C or higher.

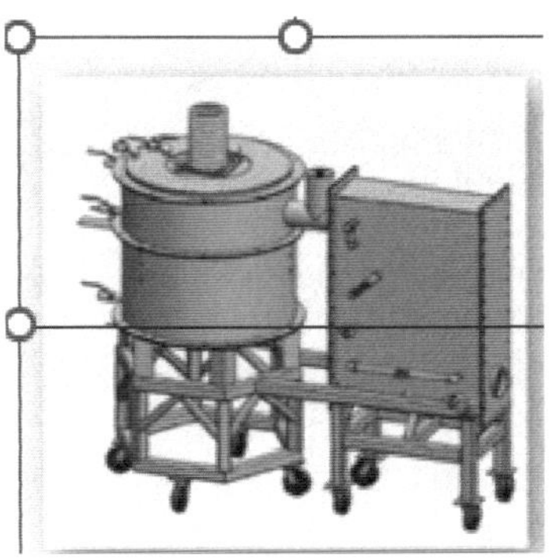

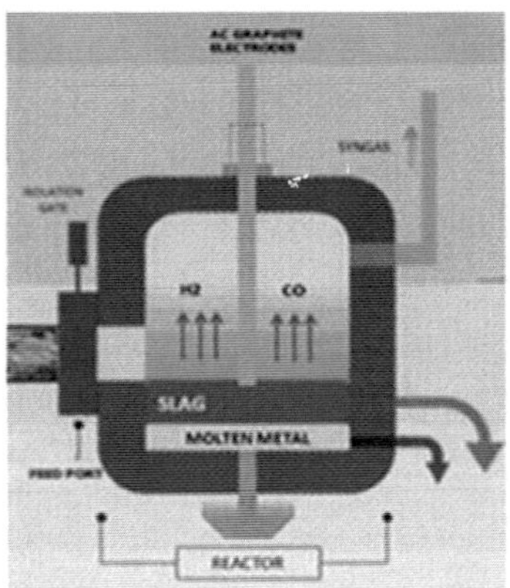

BACKGROUND

CTC (Concurrent Technologies Corporation) is building a high temperature electric arc furnace (EAF) technology demonstration unit that will, another applications, extract rare earth elements (REEs) from coal fly ash or rare earth oxides (REOs), and consequently produce slag. To fully appreciate the nuances of this gasification technology, one needs to understand the process: first, when coal fly ash and/or rare earth oxides are subjected to high temperatures, perhaps 3000°C, the carbon (organic material) will become a pure/clean syngas; the metals (iron and REEs, aluminum) will become alloys; and, everything else, the inorganics, will become slag.

REEs in order of atomic number

Name	Symbol	No.	Melting
Scandium	SC	21	1541
Yttrium	Y	39	1526
Lanthanum	La	57	920
Cerium	Ce	58	798
Praseodymium	Pr	59	931
Neodymium	Nd	60	2477
Promethium	Pm	61	1042
Samarium	Sm	62	1074
Europium	Eu	63	822
Gadolinium	Gd	64	1313
Terbium	Tb	65	1356
Dysprosium	Dy	66	1412
Holmium	Ho	67	1474
Erbium	Er	68	1529
Thulium	Tm	69	1545
Ytterbium	Yb	70	819
Lutetium	Lu	71	1662

Slag is a coarse, sand-like material that can be sold at some $18/ton to produce concrete or be sold to manufacture very expensive rock wool. To make rock wool, one needs to procure rock wool manufacturing equipment from the two suppliers in the EU (European Union).

Rock wool is used to make home insulation, sold by Owens Corning, CertainTeed, Johns Manville, among others; a different form is a hydroponic cube used in flower arranging or in greenhouses.

Generally, coal-burning utilities have ash that necessarily contains limestone slurry as part of their air quality control technologies of sulfur oxides. This alkaline slag can be used to remediate acid mine drainage (AMD) as it seeps/emerges from abandoned mines, typically found in West Virginia, Pennsylvania, and Kentucky.

The three coal refuse plants served by Robindale in western Pennsylvania will continue to generate electricity for decades since they combust reclaimed coal refuse; the water pollution benefits of removing the coal refuse piles are so great and the air pollution is minimized by using advanced flue gas desulfurization (FGD) technology. Their ash will continue to be available for thirty years, the investment life of a possible rock wool plant.

EXPERIMENTAL WORK

The ash results from the combustion of coal refuse by Seward Generation, Colver Green Energy, and Ebensburg Power. CTC would ship samples of the ash to various rock wool manufacturing plants around the United States like Johns Manville, Owens Corning, and CertainTeed (https://insulationinstitute.org/about-naima/). If the ash meets the demanding quality standards of these producers, there's a small, independent rock wool manufacturer that seeks the ash for its growth markets in New England and eastern Canada.

If there were ways to experiment with the produced ash to have it "meet spec(ifications)" that is a necessary business practice – part of continuous improvement, then modifications would be made. Perhaps, blending is an option for either the seller or buyer. This specific buyer's emerging market in New England is 100,000 tons per annum.

Like everything, economics need to be favorable. Ash is a by-product of combusting coal refuse at a cost of an estimated $18/ton to make concrete, which sells for over $30/ton.

The most valued use of slag is to become rock wool that sells for some $250/ton. There are two companies that manufacturer rock wool producing equipment – each is in the European Union (EU). Each one was name-specifically mentioned in a phone call, although Google did surface one of the two companies.

The management of Robindale Energy Services in western PA might consider installing and operating a small, 100,000 tons per year rock wool manufacturing plant.

SUMMARY

This is a unique situation since this specific plant will be operating for over thirty years; the nearby coal refuse piles will be reclaimed to keep the three plants supplied by Robindale Energy generating electricity for 650,000 residents. Neither the coal fly ash nor the bottom ash have been identified as an environmental problem. Several private companies as well as WVU, University of Kentucky, Cal Berkley, and Penn State University (PSU) have sampled the ash streams extensively and found REE concentrations in the range of 300-600 ppm levels on a consistent basis.

The environmental uses of coal refuse ash, typically mixed with limestone, which makes it alkaline, are neutralizing acid mine drainage (AMD) water/effluent that originate from underground (often abandoned) coal mines. Robindale Energy has an environmental remediation division that is very sophisticated and experienced.

Slag is used in making concrete as well as engineered fill for asphalt pavement. Robindale is very entrepreneurial and might consider manufacturing rock wool. The economics are favorable and have been revealed – the internal cost is about $18/ton, but it can sold for over $30/ton for cement. The major manufacturers of rock wool are the big players like Johns Manville, Owens Corning, among others, that probably have their own sources of slag. CTC identified an independent manufacturer of rock wool; the Pennsylvania produced slag would be used to increase the growth into under-supplied New England and eastern Canadian markets.

# # # #

FLUOR.

41st Annual International Pittsburgh Energy And Carbon Management Conference

BLUE AMMONIA AND UREA PRODUCTION VIA COAL-GASIFICATION

Ben Vivier
Ravi Ravikumar
Sunil Kumar Singhal

EMPOWERED to do it RIGHT!

PRESENTATION CONTENT

FLUOR.

INTRODUCTION

FLUOR.

- 3,500 MTPD (dry basis) coal-gasification plant with carbon capture located in the U.S.
- Two configurations shown: blue ammonia (NH_3) or blue urea ($CO(NH_2)_2$).
  - More CO2 is sequestered in the ammonia configuration.
- Sulfuric acid and Diesel Exhaust Fluid (DEF) are byproducts (urea configuration).
  - Diesel Exhaust Fluid (DEF)- used for NOx control in diesel engines.
- Facility generates its own power, importing 5 MW of electricity from the grid.

BLOCK FLOW DIAGRAM- AMMONIA CONFIGURATION

FLUOR.

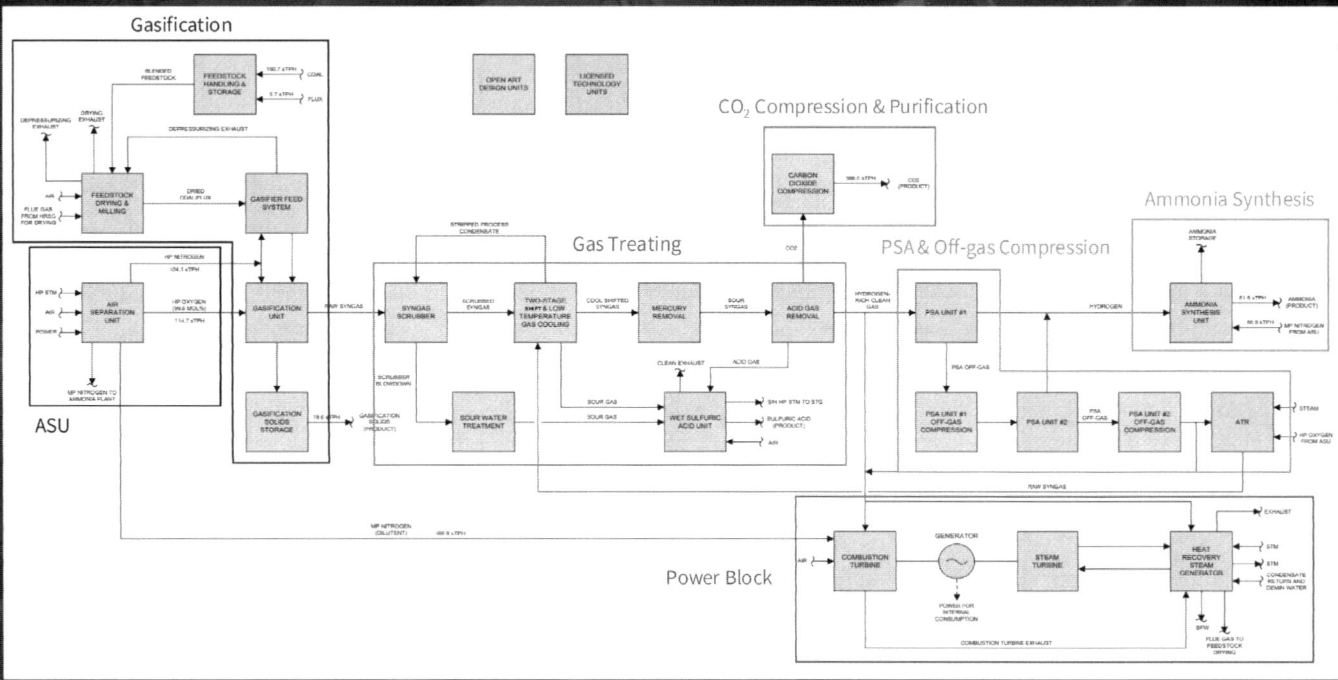

BLOCK FLOW DIAGRAM- UREA CONFIGURATION

FLUOR

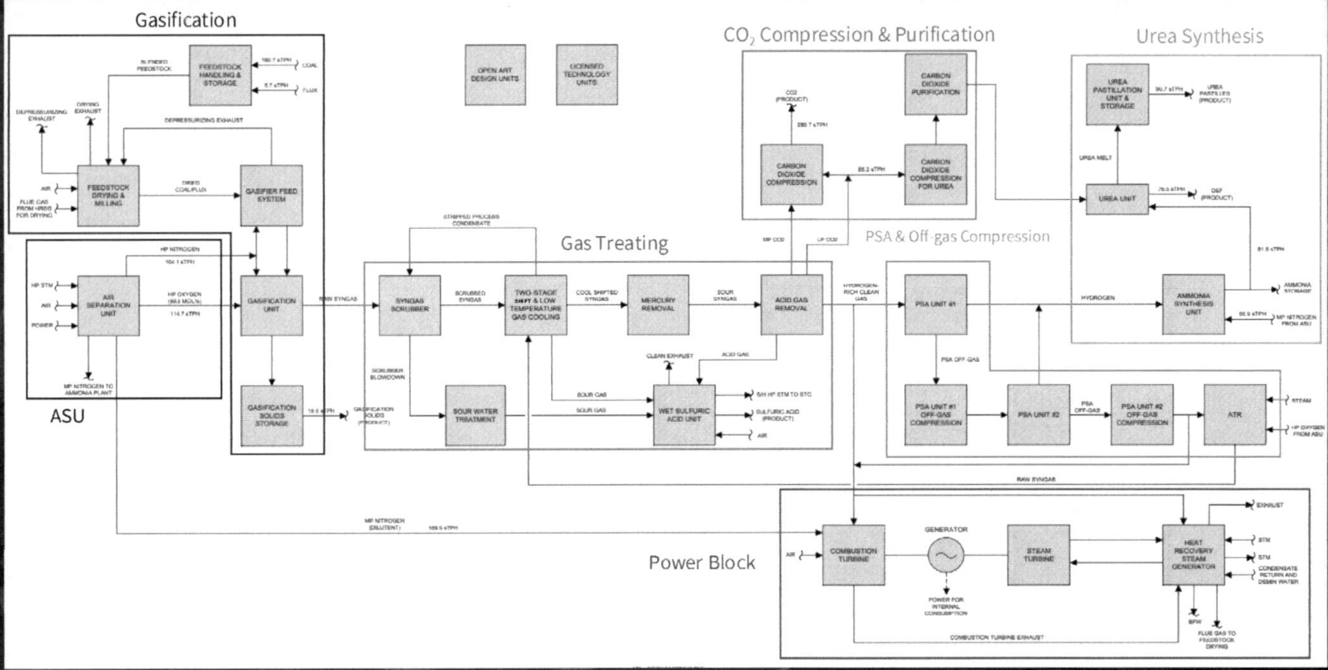

GASIFICATION

FLUOR

- 3,858 sTPD of coal and 137 sTPD of fluxant is first ground and dried.
- HP O_2 from the ASU and the feedstock are added to a two-stage gasifier.
- Gasifier produces a CO- and H_2-rich syngas.

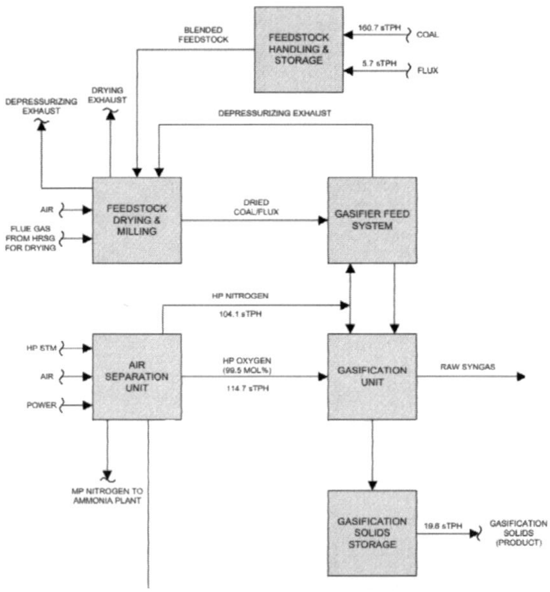

GAS TREATING

FLUOR

- Raw syngas from the gasifier is scrubbed to remove chlorides and particulates.
- CO and water vapor in scrubbed syngas is shifted to H_2 and CO_2 in two shift reactors.
- Sour Water Stripper removes H2S, NH3, and HCN from the scrubber blowdown.

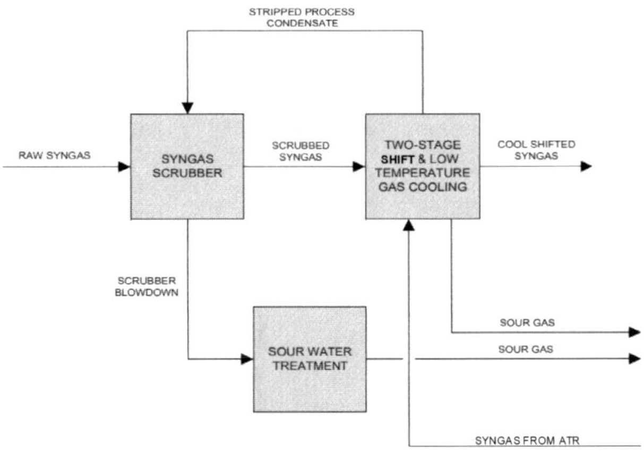

GAS TREATING

FLUOR

- Mercury and aromatics in shifted syngas is adsorbed onto an activated carbon bed.
- Acid Gas Removal Unit removes CO_2 and H_2S from the shifted syngas using Rectisol.
- Wet Sulfuric Acid Unit converts sulfur components in sour gas to sulfuric acid.

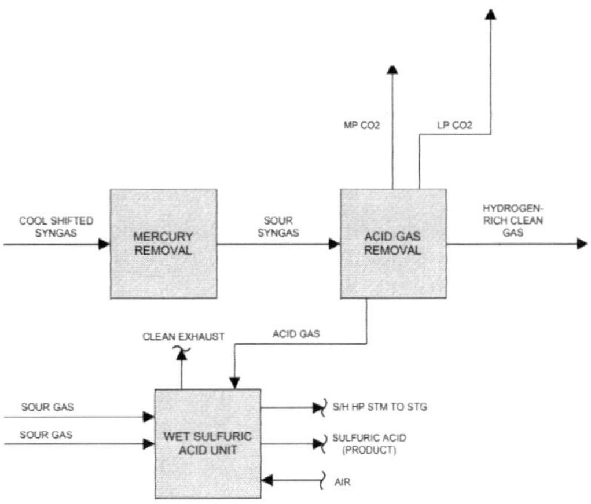

FLUOR.

CO2 COMPRESSION & PURIFICATION

- For the urea configuration, MP CO_2 and some of the LP CO_2 from the AGR Unit is compressed to 2,565 psig.
  - The remaining LP CO_2 is used for urea synthesis.
- For the ammonia configuration, all the CO_2 is compressed for sequestration.

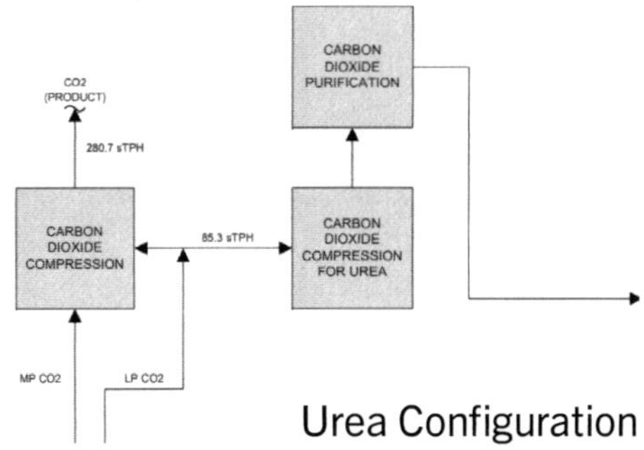

Urea Configuration

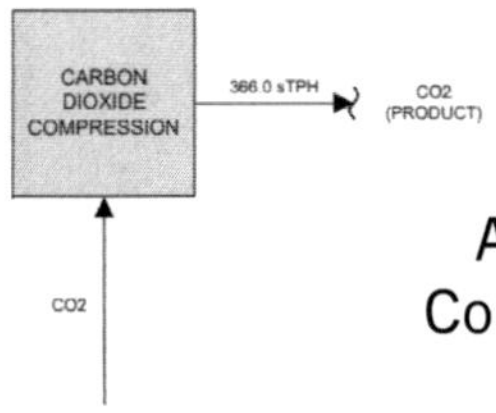

Ammonia Configuration

FLUOR.

PSA & OFF-GAS COMPRESSION

- The PSA adsorbs impurities, producing a high-purity H_2 stream.
- 76% of the PSA off-gas passes through an autothermal reformer (ATR) to generate more syngas.

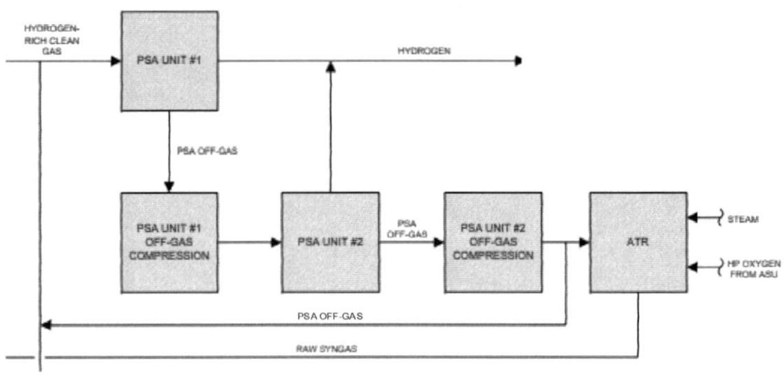

BLUE AMMONIA/UREA SYNTHESIS

FLUOR.

- MP N_2 from the ASU reacts with H_2 from the PSA to form blue NH_3.
- For the urea configuration, LP CO2 and NH3 produce blue urea.
  - 22% of the urea is concentrated and can be sold as blue Diesel Exhaust Fluid (DEF) with the required dilution.

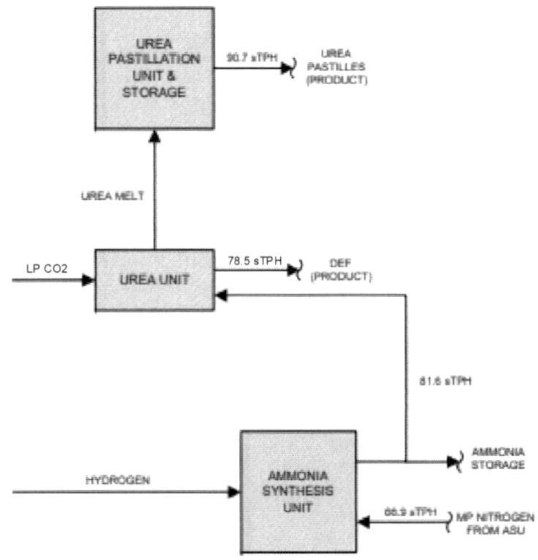

POWER BLOCK

FLUOR.

- H_2-rich clean gas from the AGR Unit and PSA off-gas are fed to the CTG.
- CTG exhaust passes through the HRSG, producing steam for the STG.

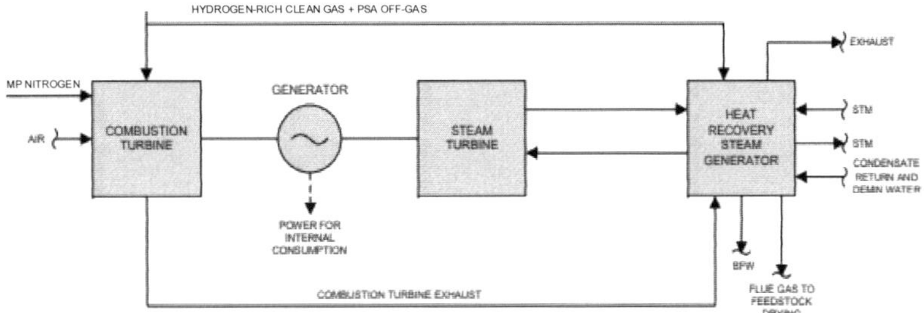

FLUOR

PLANT PERFORMANCE DATA

		Ammonia Configuration	Urea Configuration
Coal feedstock flow rate	sTPD	3,858	3,858
Product flow rate (Ammonia or Urea)	sTPD	1,968	2,184
Sulfuric Acid flow rate	sTPD	43	43
CO2 sequestration rate*	sTPD	8,784	6,744
Carbon Capture	%	95	95
Power Requirement	MW	222	228
Power Generation	MW	222	223
Power Imported	MW	0	5

*Meets green ammonia carbon intensity (<0.3 kg CO_2e/kg NH_3)

FLUOR

PLANT ECONOMICS

	Ammonia Configuration	Urea Configuration
Cost of Production (MM USD/yr)	119	164
Profit (MM USD/yr)	433	411
Simple Payback Period (yrs)	12	13

- 45Q tax credit for carbon capture & sequestration (85 USD/MT) greatly reduces operating costs.
- Blue ammonia product meets green ammonia carbon intensity (0.3 kg CO_2e/kg NH_3), so the ammonia price is set at 931 USD/MT.
  - The carbon intensity excludes upstream emissions.
- If all the urea produced in the facility were used for DEF, the profit would be 585 MM USD/yr with a simple payback period of 9 years.

SOURCES

FLUOR

- [1] L. Sierra, "WSA Technology a competitive solution for sulfur management." Available: https://refiningcommunity.com/wp-content/uploads/2017/06/WSA-technology-a-competitive-solution-for-sulfur-management-Llorens-Haldor-Topsoe-SRU-Galveston-2018.pdf
- [2] P. Baboo, "The Snamprogetti Urea Process," *UreaKnowHow.com*, Jan. 2021. https://ureaknowhow.com/wp-content/uploads/2021/01/2021-01-Baboo-Snamprogetti-Urea-Process.pdf
- [3] "Engineering Economic Analysis Guide: Liquid Fuels Technologies," Dec. 2015. https://www.eia.gov/outlooks/documentation/workshops/pdf/eeaguide_12082015.pdf
- [4] L. Ryan, "Ammonia's Potential Role in a Low-Carbon Economy," *Congressional Research Service*, Dec. 07, 2022. https://crsreports.congress.gov/product/pdf/IF/IF12273 (accessed Sep. 17, 2024).
- [5] A. Jones, "The Section 45Q Tax Credit for Carbon Sequestration," Aug. 25, 2023. https://crsreports.congress.gov/product/pdf/IF/IF11455
- [6] "Interactive: Ammonia price chart," *www.spglobal.com*, Jul. 20, 2023. https://www.spglobal.com/commodityinsights/en/market-insights/latest-news/energy-transition/051023-interactive-ammonia-price-chart-natural-gas-feedstock-europe-usgc-black-sea
- [7] Green Hydrogen Organisation, "GH2 Green Hydrogen Standard Fact Sheet." https://gh2.org/sites/default/files/2023-03/GH2_Standard_Fact%20Sheet_v2_DIGITAL.pdf#:~:text=Green%20hydrogen%20is%20hydrogen%20produced%20through%20the%20electrolysis,taken%20as%20an%20average%20over%20a%2012-month%20period%29.

EMPOWERED to do it RIGHT!
AWARENESS. COLLABORATION. ESCALATION

FLUOR

QUESTIONS?

EMPOWERED to do it RIGHT!
AWARENESS. COLLABORATION. ESCALATION

BUILDING AN ENERGY SYSTEM BASED ON THAR COAL

2024 International Pittsburgh Coal Conference
Pittsburgh (Virtual), October 7 – 9, 2024

Dr. Farid A. Malik
FCC University, Lahore – Pakistan

Engr. Abid Aziz
Pak Motors (Pvt.) Ltd., Pakistan

Table of Contents

ABSTRACT

- While the world moves away from Coal, Pakistan is forced to use its indigenous Black Gold to cut down its import bill. An energy system based on imported fuel is un-sustainable and insecure.
- The country's import bill is around $25.2 billion which includes imported Coal worth $2.0 billion and LNG $2.5 billion. The huge Coal reserve (175 billion tons) at Thar remain grossly under-utilized. Mining challenges have been overcome. Two mines are now functional, each producing 7.6 million tons per year. Power generation has started with linkage to the main transmission grid. Successful gasification has also been achieved in collaboration with experts in South Africa.
- Currently efforts are underway to start large scale gasification to build a Coal based affordable energy system for the country by application of Clean Coal technologies. Multiple products are planned that include SNG, Urea, Diesel, etc.

2

Details of Thar Coal Deposit

- Total coal resources of Thar Coal-field are estimated at 175 billion tons; lignite quality, high percentage of moisture, low in ash & Sulphur.
- Of 12 blocks of Thar coal, 04 have been allotted for coal mining, gasification & power generation.
- First open-pit mine in Thar Block-II is completed by Singh Engro Coal Mining Company for 3.8 million tons per year coal production and its upgradation to 7.6 mtpa is underway. Power generation units are also synchronized to national grid for 660 MW.
- One Coal-To-Liquid (CTL) Project for 7.8 MTPA approved by JCC on CPEC.

3

- Thar vs North Dakota Coal Deposits

Thar Coal						
SIZE	CV Btu/lb	Composition				
		Moisture	Ash	Volatile Matter	Fixed Carbon	Sulphur
175 B	8000	43.24	5.5	23.42	20.5	1.12
North Dakota Coal						
25 B	6500	30	10	34.86	40.39	0.99

4

DAKOTA GASIFICATION COMPANY SYNFUELS PLANT

•18,000 tons of lignite coal is used daily

•6 million gallons of water are used to produce steam using a closed loop system

•170 million cubic feet of natural gas is produced daily

•Raw gas is cleaned and then passed over nickel catalyst causing CO and remaining CO_2 to react with H_2 to form CH_4

•Gas is cooled dried and compressed before entering the pipeline

•Heating value of gas is 975 Btu per cubic feet (Sui gas 920 Avg)

•Other byproducts: Ammonium Sulphate, Anhydrous Ammonia, Carbon Dioxide, Dephenolized Cresylic Acids, Diesel Exhaust Fuels, Krypton and Xenon gases, Liquid Nitrogen, Naphtha, Phenol, Tar Oil, Urea

5

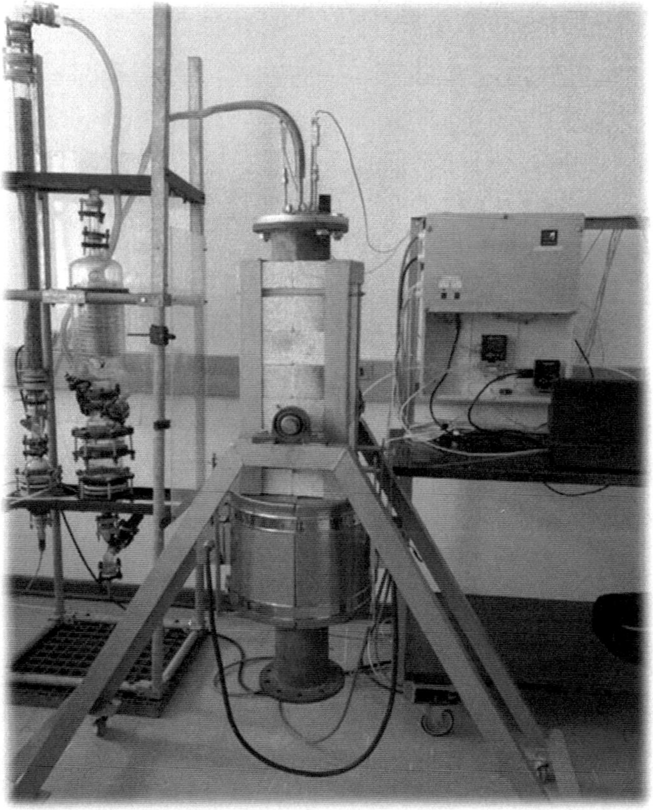

Gasifier used at Nelson Mandela University, South Africa

(Fixed bed atmospheric pressure, air blown)

6

SLAG-LIQUID PROFILE AND OPERATING GASIFICATION WINDOW

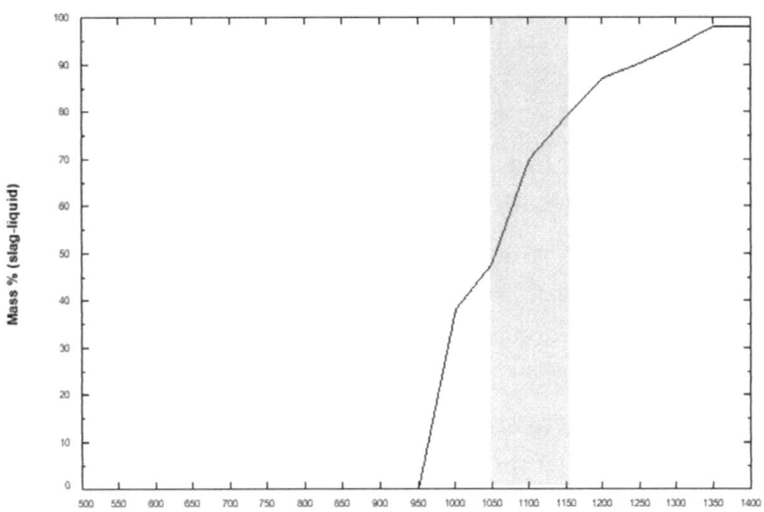

In summary, this coal (based on the samples as received) is highly suitable for fixed bed gasification purposes. No concerns or problems identified. The expected PURE GAS YIELD is between 1550-1600 kNm^3 per ton DAF coal. The operating temperature of gasification is predicted as 1050-1200°C.

7

THAR COAL DEVELOPMENT PUBLICATIONS
DR FARID A. MALIK

- May 2005 - Proper Exploitation of Thar Coal-DAWN
- Oct 2010 - Thar Coal Mining Challenges-IPCC, Istanbul
- Sep 2011 - Development of Thar Coal Putting the Pieces Together IPCC, Pittsburgh
- May 2012 - Applicability of the UCG process; Dig or Skip, UCG Conference, London
- Oct 2012 - Thar Coal: Dig or Skip, IPCC, Pittsburgh
- Sept 2013 - Reaching for bulk samples through Aquifers at Thar Coal, IPCC, Beijing
- Oct 2014 - Producing Urea from Thar Coal Deposit, IPCC, Pittsburgh
- Oct 2015 - Use of Coal in the Energy Mix of Pakistan (Keynote Address), IPCC, Pittsburgh
- Aug 2016 - The future of Coal IPCC, Cape Town
- Sept 2017 - Replacing SNG with LNG in the Energy Mix of Pakistan, IPCC, Pittsburgh
- Oct 2018 - Producing SNG from Thar Coal in Pakistan, IPCC, Xuzhou, China
- Sept 2019 - Building Coal Based Energy System in Pakistan, IPCC, Pittsburgh
- Sep 2020 - Applications of Clean Coal Technologies for development of Thar Coal in Pakistan, IPCC, Pittsburgh
- Sept 2021 - An Integrated Energy System for Thar Coal in Pakistan, IPCC, Pittsburgh
- Sept 2022 - Road Map for Producing SNG from Thar, Coal in Pakistan
- Oct 2023 – Experience of SNG Production from Thar Coal in Pakistan
- Oct 2024 – Building an Energy System Based on Thar Coal

8

BUILDING AN ENERGY SYSTEM BASED ON THAR COAL

CONCLUSIONS

1. Pakistan has huge coal reserves (184 B.ton) that can be utilized by application of appropriate technologies starting with Gasification
2. As the country has an extensive gas distribution network it will continue to cater for the energy needs
3. Imported LNG is a short-term fix, SNG and new gas discoveries have to be brought on-line to meet the shortfall (2000 mmcfd)
4. With the current LNG prices above $ 20/ MMBTU, SNG from indigenous coal is viable.
5. Thar is the future energy resource for the country with multiple products option (SNG, Urea, CTL/FTD etc.)
6. As Gasification of Thar Coal has been established, it may prove to be an important component of Pakistan’s energy mix.
7. Way forward is large scale Gasification of Thar Coal to form basis of Clean Coal Energy System.
8. Pakistan seeks technology partners for large scale Gasification of Thar Coal.

9

Machine Learning Enhanced LIBS to Measure and Process Biofuels and Waste Coal for Gasifier Improved Operation[1]

Ozan Karadut[a] Zheng Yao[a], Carlos Romero[a], Robert De Saro[b], Joe Craparo[b], Larry Mulligan[b], Johan Van Dyk[c], Asmara Soomro[c], Alexander Koutsostamatis[c], Steve Goff[d 1]

[a] Energy Research Center, Lehigh University, 117 ATLSS Dr., Bethlehem, PA 18015, USA

[b] Energy Research Company, 400 Leland Ave, Plainfield, NJ 07062, USA

[c] GTI Energy, 1700 S Mount Prospect Road Des Plaines, IL 60018, USA

[d] SpG Consulting, USA

Abstract:

Waste coal, coupled with biomass, as feedstock in gasifiers that produce hydrogen faces some challenges, such as the large heterogenic nature of the components, the slagging operation of the gasifier due to the ash content and composition of the coal waste and biomass blend, and the need to maintain tight control of the feedstock specifications. A machine learning (ML) enhanced Laser Induced Breakdown Spectroscopy (LIBS) concept has been investigated to measure the characteristics of the feedstock going into the gasifier in situ and in near real-time, resulting in time-sensitive fuel data that gasifier operators can use in feedback or feedforward control schemes to maximize performance and mitigate the deleterious impact of feedstock ash properties on gasifier operation. LIBS is capable of measuring the elemental spectroscopic emissions of the feedstock while it is flowing over a belt. The LIBS spectral is then processed by ML algorithms, which produce data to be used by either plant operators or by a feedback or feedforward automatic control system.

1 Introduction:

[1] This paper was presented at 41st annual international Pittsburgh energy and carbon management conference

Coal waste is the material left over from mining and typically represents 40% of that which is mined. Utilizing waste coal as a feedstock in gasification processes presents an opportunity to extract energy while avoiding many of the environmental downsides associated with coal combustion. In particular, the co-gasification of coal waste with biomass has emerged as a promising approach to hydrogen production, reducing carbon emissions while making use of otherwise discarded materials.

Laser-Induced Breakdown Spectroscopy is a cutting-edge technique for elemental analysis that enables in-situ and in near real-time, resulting in immediate and time-sensitive data for material characterization and it operates by focusing a high-energy laser onto a sample. When the laser hits the material, it generates a small plasma, that contains excited atoms and ions from the sample. This plasma emits light at specific wavelengths that correspond to the elements present in the sample. This light is collected via a fiber optic cable and sent to a spectrometer, which measures the emitted light in terms of both intensity and wavelength to determine material composition. Traditionally, LIBS has been used for static laboratory analysis; however, in this concept, LIBS measures the elemental spectroscopic emissions of the feedstock while it is flowing over a belt or being transported in a pneumatic line.

The spectrum is then processed by ML algorithms which generate the final data to be used by either the plant operators or by a feedback or feedforward automatic control system. This integration of ML with LIBS supports real-time decision-making and enhances accuracy, as the models continuously refine their precision through ongoing training on new data. Furthermore, this concept is scalable for industrial applications because the ML models can handle large amounts of data after being trained in the ML. By utilizing these predictions, plant operators can optimize gasification processes, leading to greater efficiency and sustainability.

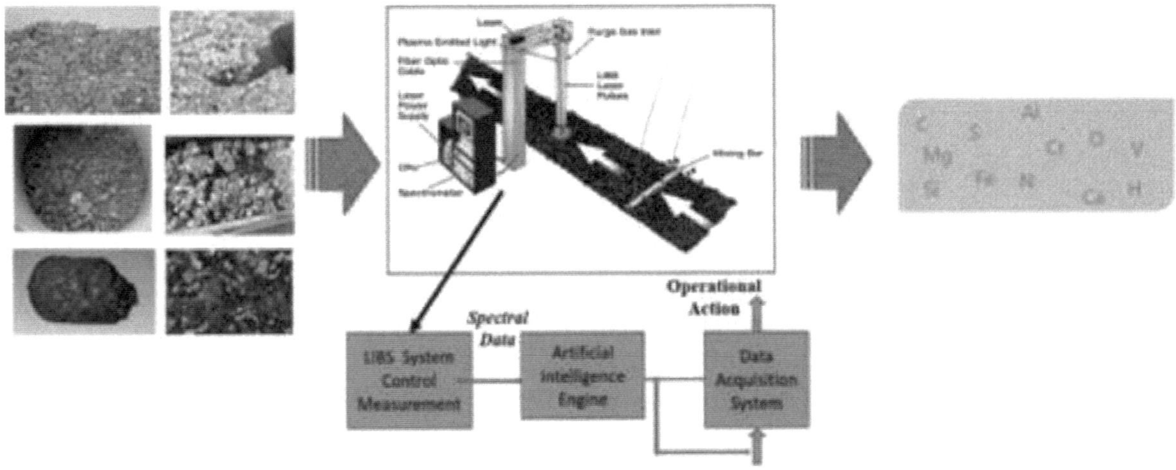

Figure 1: LIBS + ML Concept

In short, the ML enhances LIBS by making the system smarter and more efficient. The feedstock material matrix for this study consists of both pure and blended samples. The pure samples include Biomass Wood Pellet from GTI, Vineyard Biomass, Waste Coal from GTI, and Dominion Waste Coal. In addition to these pure materials, various blended samples have been prepared to explore the effects of combining materials in different ratios and particle sizes. To improve the accuracy and performance of machine learning models, each sample group was sent to a certified laboratory for comprehensive analysis to obtain the target variables, which are essential for model training and validation. In this study, various machine learning algorithms were applied to analyze the LIBS data such as Linear Regression, Support Vector Machines, Random Forest, Neural Networks, Gradient Boosting, and Deep Learning through Python libraries including scikit-learn, TensorFlow, and Keras. Among the models tested, Random Forest and Gradient Boosting demonstrated the best performance on the LIBS data, highlighting their effectiveness for this application.

2 Feedstocks Description and Processing

2.1 Feedstock Material Matrix:

Table 1 indicates the feedstock material matrix, presenting five distinct materials used in its composition. The matrix is organized into two major sample groups: pure and blended materials, each with different mixing ratios among the ingredients. There are four different feedstocks which are GTI Biomass Wood Pellet, Vineyard Biomass, GTI Waste Coal, and Dominion Waste Coal.

In the blend materials group, various combinations of these feedstocks are formulated with two size-based variations for each blend: one with an average particle size of 2 mm, and another with an average size of 500 microns. Blend-1's ingredients are 75% GTI Biomass Wood Pellet and 25% Dominion Waste Coal. Blend-2 consists of 95% GTI Biomass Wood Pellets and 5% GTI Waste Coal, and Blend-3 includes 99% GTI Biomass Wood Pellets and 1% GTI Waste Coal. Unlike other blends, Blend-4 is composed of Blend-1 and Calcium Carbonate (Limestone), creating a different formulation within feedstock sample groups. Feedstock material details are shown in Table 1.

Table 1: Feedstock Material Matrix

Sample Blending Ratio					
Sample Name	Ingredient 1		Ingredient 2		Blend size
	Material Name	Blending Ratio (%)	Material Name	Blending Ratio	
Pure-1	GTI Biomass Wood Pellet	100.00%	-	0.00%	
Pure-2	Vineyard Biomass	100.00%	-	0.00%	
Pure-3	GTI Waste Coal	100.00%	-	0.00%	
Pure-4	Dominion Waste Coal	100.00%	-	0.00%	
Blend-1	GTI Biomass Wood Pellet	75.00%	Dominion Waste Coal	25.00%	1-3 mm
Blend-1	GTI Biomass Wood Pellet	75.00%	Dominion Waste Coal	25.00%	500 microns
Blend-2	GTI Biomass Wood Pellet	95.00%	GTI Waste Coal	5.00%	1-3 mm
Blend-2	GTI Biomass Wood Pellet	95.00%	GTI Waste Coal	5.00%	500 microns

Blend-3	GTI Biomass Wood Pellet	99.00%	GTI Waste Coal	1.00%	1-3 mm
Blend-3	GTI Biomass Wood Pellet	99.00%	GTI Waste Coal	1.00%	500 microns
Blend-4	Blend-1	93.50%	Limestone	6.50%	1-3 mm
Blend-4	Blend-1	93.50%	Limestone	6.50%	500 microns

2.2 Feedstock Samples:

Two major feedstock types are used for this project. The first feedstock type is waste coal, a material left over from mining that typically represents 40% of mined material. Two types of waste coal are used: Dominion Waste Coal and GTI Waste Coal. The second type of feedstock is biomass, a renewable organic material derived from wood, plants, and agricultural waste. Two types of biomasses are used: GTI Biomass Wood Pellet and GTI Vineyard biomass. Additionally, limestone, a naturally occurring sedimentary rock primarily includes calcium carbonate, is used in creating Blend-4.

2.3 Woodchips Preparation:

The wood pellets were initially processed using a shredder machine to reduce them into smaller particles, resulting in wood chips. After shredding, the material was separated by using a sieve shaker to classify based on dimensions. This sieving process took 5 minutes per 200 grams of woodchips to ensure consistent size separation.

Following sieving, the woodchips were categorized into two main size groups. The first group, with an average particle size of 500 microns, contained finer particles, while the second group, averaging 2 mm, consisted of larger particles. By controlling particle size, we aim to understand better how these dimensions impact the LIBS and ML systems.

2.4 Coal Preparation:

Unlike the biomass, coal couldn't be processed with a shredder machine because of the particle's hardness. Instead, the coal samples were crushed using a hand mortar and a small crusher machine. After the crushing process, the coal samples were sieved using a shaker and classified into two main groups the same as woodchips.

Any coal particles larger than 2.8 mm were reprocessed using the same procedure to ensure the blends achieved the desired particle size and consistency for further analysis. This meticulous particle size classification has critical importance for the blends because it ensures homogeneity and improves the accuracy of the subsequent LIBS analysis.

2.5 Blend Preparation:

A concrete mixer was employed to combine the ingredients to reach homogeneity for each blend. The feedstocks were added to the mixer, and then the machine was operated for 30 minutes per batch. To enhance the homogeneity of the blends, the mixer was worked with an angle. In addition, during the mixing process, the direction of the mixer was changed periodically. This step further improved the uniformity of the blends, ensuring that the final mixture was well-distributed and consistent in composition.

2.6 Laboratory Analysis Results:

To improve the accuracy and performance of the machine learning models, each sample group was sent to a certified laboratory. Samples were selected from different levels of the buckets to maximize the representation of the sample group. Also, each sample group was analyzed twice to enhance the machine-learning capabilities by providing a more reliable and consistent data set.

The analyses focused on proximate, ultimate, mineral composition, and the fusion temperature of ash, key factors that play a significant role in material elemental composition prediction and gasification. These analysis results are key output parameters for training and validating the models.

2.7 Representative LIBS Spectrum:

A representative LIBS spectrum consists of emission peaks corresponding to the specific elements present in the sample. Each element emits light at unique wavelengths, enabling the elemental identification by analyzing these peaks. LIBS captures data over a wide spectral range, giving a comprehensive view of the sample's elemental composition. Furthermore, the intensity of these peaks is proportional to the concentration of each element, meaning the higher the intensity, the greater the amount of that element in the sample. Measuring wavelength intensity helps to accurately predict material composition in real-time.

LIBS data sets are creating input values, while certificated laboratory analysis results are providing the target values. Machine learning algorithms are trained on these datasets to develop models capable of making accurate predictions. Therefore, different machine learning algorithms were implemented to reach the optimum point concerning Root Mean Square Error (RMSE) or R^2 value, as well as computational cost. In the following sections, implemented machine learning techniques and results are briefly described.

3 Machine Learning Methods:

Seven distinct machine learning algorithms were implemented to predict targeted variables, by considering minimizing RMSE and maximizing R2 value. For this study, Python and its powerful libraries, such as scikit-learn, Tensor Flow, and Keras were used to run Linear Regression, Support Vector Machine, Random Forest Regression, Neural Network, Gradient Boosting, Deep Learning Sequential API, and Ensemble Methods which consist of three major machine learning methods; Support Vector Machine, Random Forest Regression, and Neural

Network method. Seven distinct machine learning methods were applied to LIBS's measurement data set for Al, Ca, Fe, Mg, Na, Si, Ti, K, and C, to predict laboratory analysis results which included the heating value of material, ash composition, carbon composition, and sulfur composition, as well as SiO_2, Al_2O_3, Fe_2O_3, CaO, MgO, K_2O, Na_2O content. Different input parameters were used to predict each output; for instance, carbon content was based on the C, Ti, Al, O, and Mg elemental spectroscopic emission intensities, and ash content was based on Al, Ca, Fe, Mg, Na, Si, Ti, and K elemental spectroscopic emission intensities. Different hyperparameters were applied to maximize the R^2 value and minimize RMSE for the training and test data sets. After the tuning operations, Random Forest Regression and Gradient Boosting methods look like the best way to predict target variables.

3.1 Random Forest Regression:

Random Forest Regression is a robust machine learning algorithm that leverages ensemble learning to enhance the reliability and accuracy of predictions by averaging different predictions. By merging distinct decision trees, each trained on a unique subset of the data, the model reduces overfitting and improves generalization. Through this aggregation, Random Forest lowers variance and boosts model accuracy. Each tree is trained on a random data subset, creating diverse models that are less prone to overfitting and better at generalizing to unseen data.

However, it's important to note that Random Forest Regression can be computationally expensive, especially with large datasets, requiring substantial processing power and memory resources. Despite this trade-off, Random Forest Regression remains a powerful and versatile tool for a wide range of regression problems.

3.2 Gradient Boosting Method:

Gradient Boosting is an ensemble technique that builds models sequentially, where each new model corrects the errors of its predecessors. This method is an additive model, which combines multiple weak learners like typically decision trees, to create a strong predictive machine learning model. Gradient Descent Optimization is used to decrease the error by adjusting the model weights at each step, ensuring continuous performance improvement. One of Gradient Boosting's key advantages is its capacity to capture non-linear relationships in the data set, making it particularly effective for complex datasets. Its ability to incrementally enhance prediction accuracy renders Gradient Boosting an excellent choice for datasets with intricate patterns and non-linearities.

4 Machine Learning Results:

The following sections explain the machine learning results for five target elements: Moisture Ash-Free Heating Value, Ash % (Dry Basis), Carbon, Sulfur, and SiO_2. For each target element, there are two types of graphs. The first graph uses the actual results, while the second one uses the normalized results among different methods. The left y-axis and bar charts represent the RMSE, and the right y-axis and line graphs indicate the R^2 value. The x-axis shows the machine learning methods that were implemented. The primary purpose of this different machine learning algorithm is minimizing RMSE and maximizing R^2 value. By utilizing both RMSE and R^2 scores of the training and testing datasets, the performance of each model can be effectively assessed, and overfitting can be avoided. Therefore, graphs were organized to show overfitting possibility. When the lines and bar charts converge at the same point or within the same region, it indicates that overfitting is likely avoided.

4.1 Moisture Ash Free Heating Value:

The first target variable is Moisture Ash Free Heating Value. Random Forest Regression and Gradient Boosting methods show the best performance for the target value among the implemented machine learning algorithms as well R2 values both for training and testing data sets exceeding 0.95. On the contrary, Linear regression and Support Vector machines are not robust enough to capture the non-linearities so they give the worst results. In addition, there is an overfitting problem for neural networks and deep learning sequential API. Because there are large discrepancies between the training and testing data set for R2 and RMSE values. In addition, deep learning sequential API and neural network methods yield very similar results due to their structural similarity.

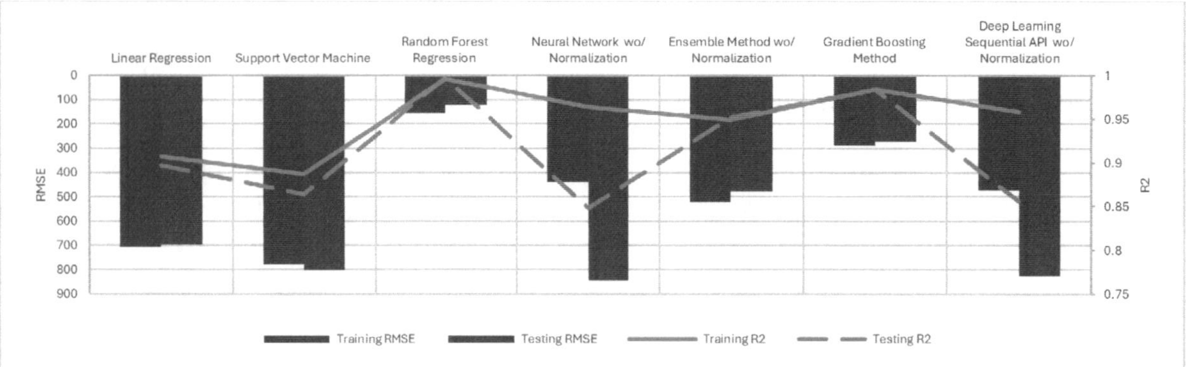

Figure 2: Comparison of Machine Learning Model Performance: Moisture Ash Free Heating Value

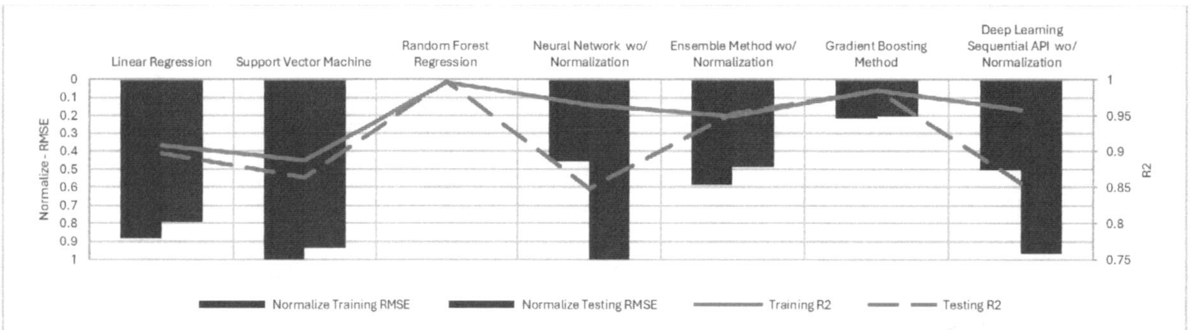

Figure 3: Comparison of Machine Learning Normalization Model Performance: Moisture Ash Free Heating Value

4.2 Ash (Dry Basis)

Ash content on a dry basis is the second target variable. The scores of the training and testing data set are very close to each other for all machine learning techniques except Deep Learning Sequential API, so overfitting is not considered as a problem. However, a point of concern arises with the Deep Learning Sequential API. There is a significant difference between the training and testing RMSE values, suggesting potential overfitting. This indicates that the model may be learning the training data too well, which could lead to poor generalization of unseen data and overfitting. Random Forest Regression emerges as the best method for predicting ash content among all the methods. Furthermore, the Neural Network, Ensemble Method, and Gradient Boosting Method produce comparable results, with R2 values exceeding 0.94 for all three. This high level of performance indicates that these methods are robust in modeling the ash content, providing reliable predictions. Overall, while Random Forest stands out as the top performer, the close performance of the other three approaches suggests that they are also strong candidates for further exploration.

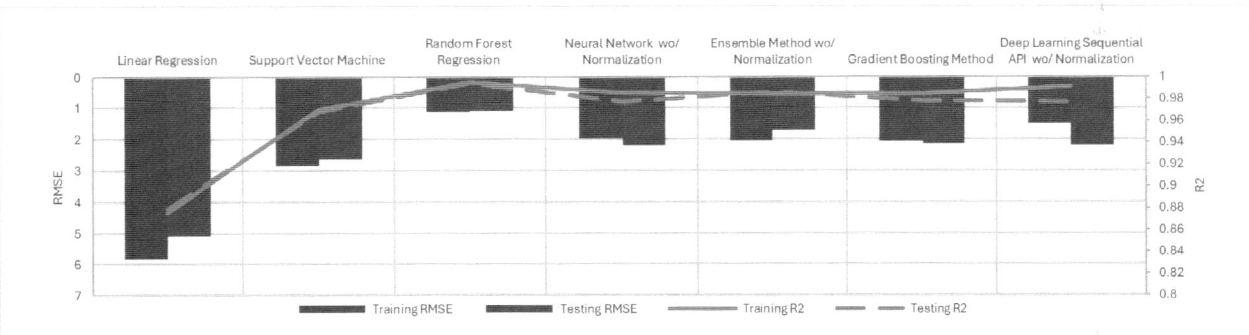

Figure 4: Comparison of Machine Learning Model Performance: Ash Dry Basis

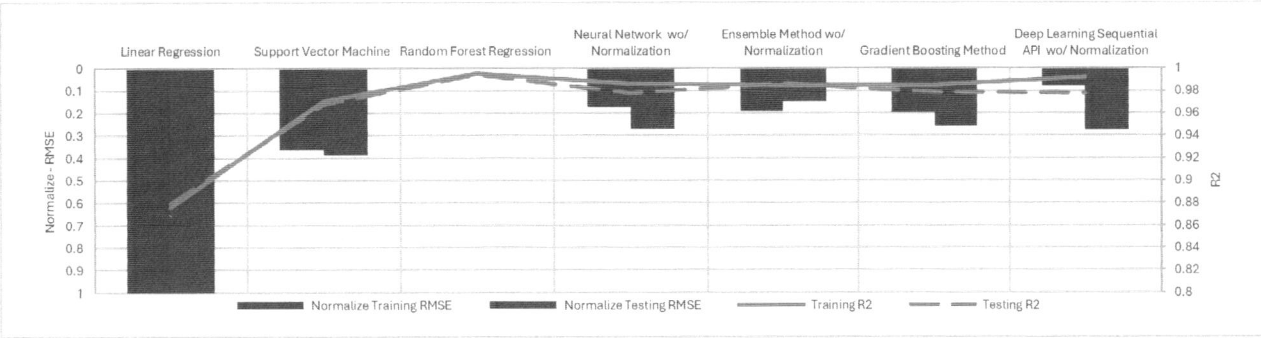

Figure 5: Comparison of Machine Learning Normalization Model Performance: Ash Dry Basis

4.3 Carbon:

The Gradient Boosting Method is the best way to predict the carbon content of the feedstock materials because it is achieving lowest training and testing RMSE values and the R2 value is very close to 1. However, unlike the other material composition prediction, Random Forest regression shows an overfitting problem due to the RMSE value differences between the training and testing datasets. Expanding the data set can help to solve this issue and improve generalization. On the contrary, Linear Regression and Support Vector Machine techniques performed the worst demonstration of the need for more complex models to capture the non-linearities of the data set. Overall, Gradient Boosting has proven to be the best fit for the carbon. Predictive capabilities can be enhanced with further tuning and a larger data set with an optimistic approach.

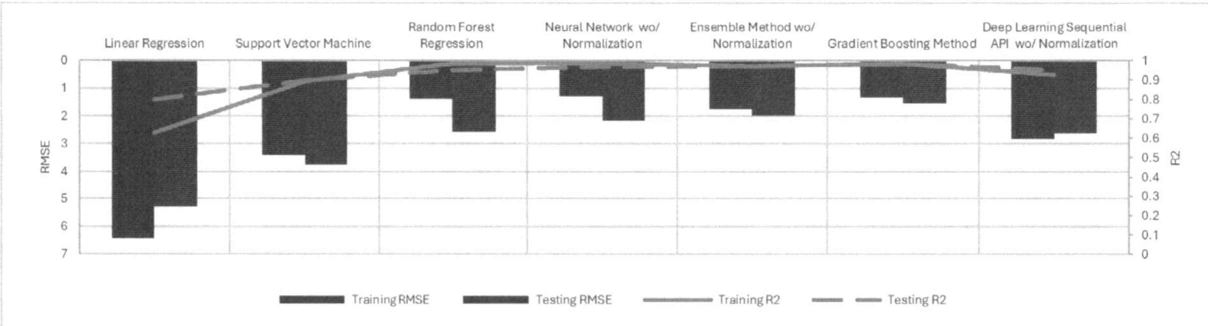

Figure 6: Comparison of Machine Learning Model Performance: Carbon

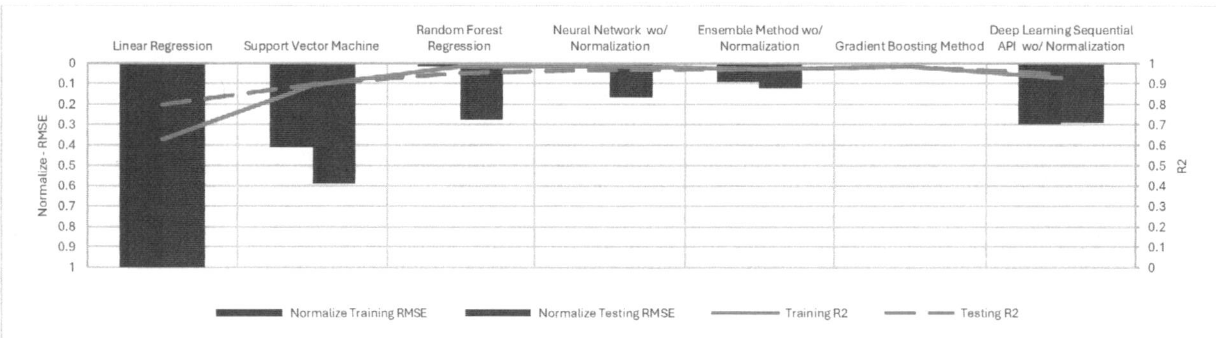

Figure 7: Comparison of Machine Learning Normalization Model Performance: Carbon

4.4 Sulfur

Sulfur content is generally quite low when compared to other elemental components. This characteristic results in a smaller range for the RMSE when sulfur is analyzed, making it a unique challenge. Therefore, normalization of the RMSE values is a good way to show the performing differences of the distinct techniques. The Random Forest regression methods show the most accurate prediction method among the implemented techniques based on the RMSE and R2 values. In addition, there is overfitting problem on the Neural Network method because of the testing and training data set RMSE and R2 value differences. The Gradient Boosting method is the second-best method to estimate sulfur content of the materials.

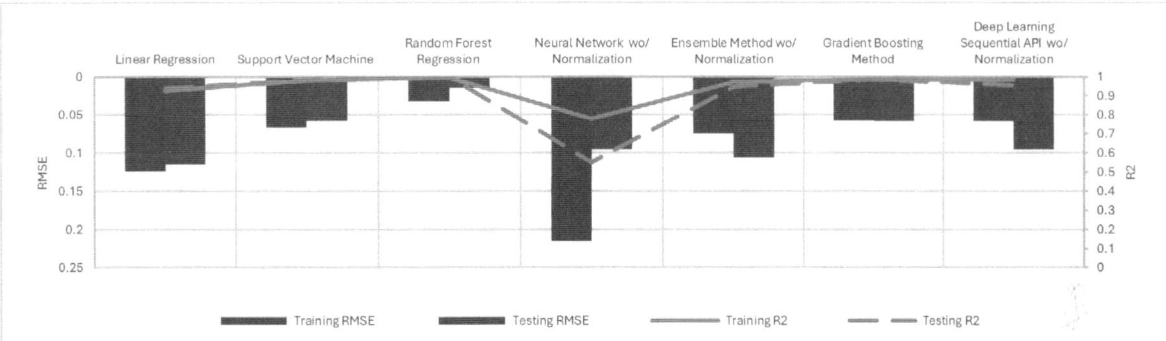

Figure 8: Comparison of Machine Learning Model Performance: Sulfur

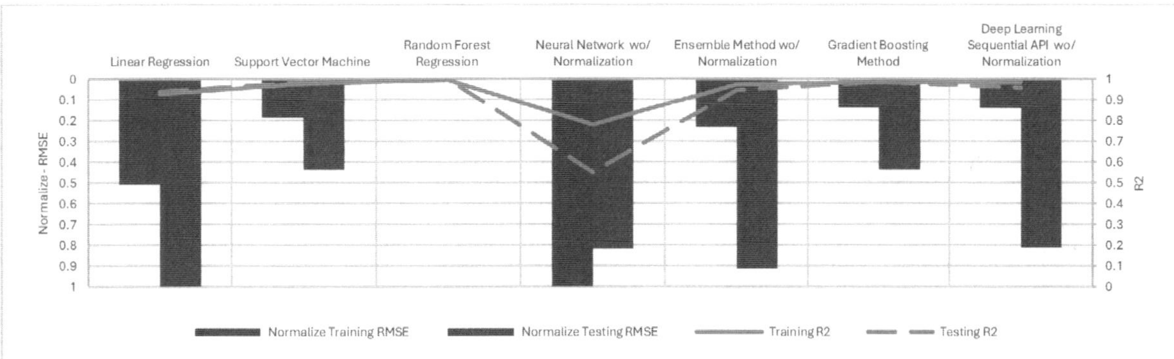

Figure 9: Comparison of Machine Learning Normalization Model Performance: Sulfur

4.5 SiO_2

The Random Forest regression method is the most effective approach for estimating SiO_2 content; however, there may be overfitting due to the difference between the RMSE values for the training and testing datasets. Additionally, the Neural Network method produces good results but shows significant differences between the RMSE values of the training and testing datasets. This indicates a need for further tuning and an increase in dataset size, which could improve the model's generalization ability. In contrast, Linear Regression and Support Vector Machine methods are insufficient for capturing the complex relationships in SiO_2 data.

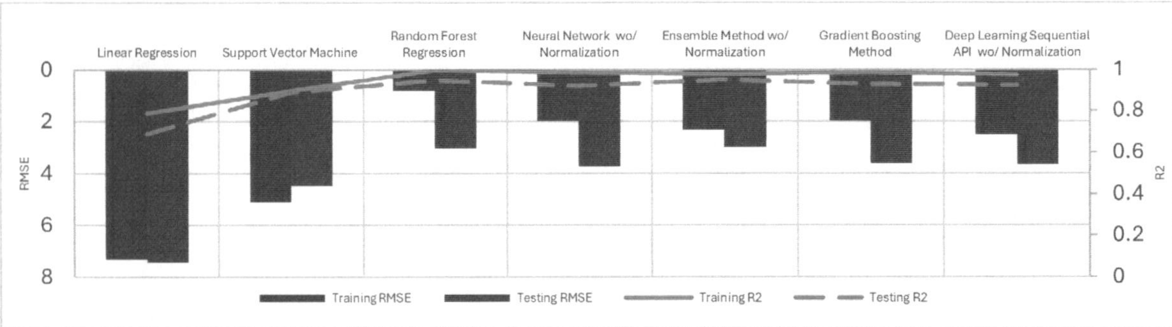

Figure 10: Comparison of Machine Learning Model Performance: SiO_2

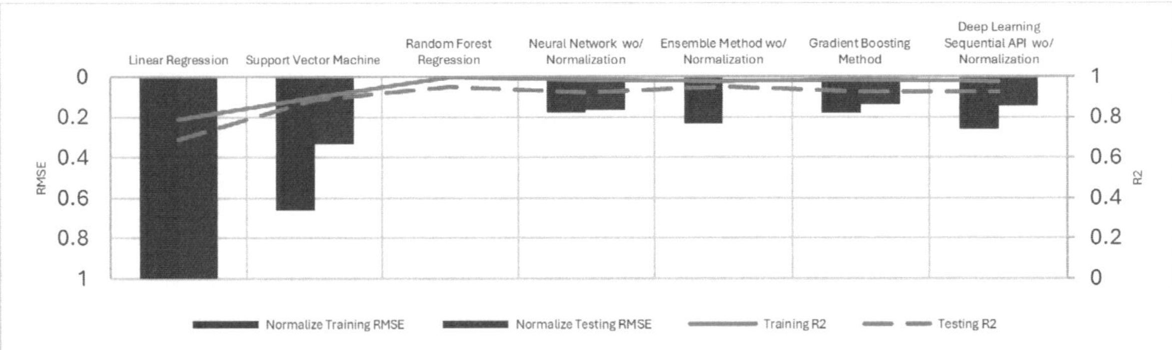

Figure 11: Comparison of Machine Learning Normalization Model Performance: SiO_2

5 Conclusion

To sum up, four different pure materials which are GTI Biomass Wood Pellets, Vineyard Biomass, Waste Coal from GTI, and Dominion Waste Coal were obtained and processed to prepare two different sizes of the blends. After collecting samples, the initial phase involved collecting and analyzing data through LIBS and ML systems.

During ML, advanced regression techniques emerged to predict materials' composition. Among the implemented models, Random Forest regression and Gradient Boosting proved to be the most effective. These two regression techniques consistently produced the highest R2 values and lowest RMSE for the LIBS system working with machine learning.

Exceptional results were achieved for specific components: an impressive R^2 value of 0.99899 for sulfur content using Gradient Boosting, and an R^2 of 0.9843 for heating value with Random Forest Regression. These outcomes highlight the models' accuracy in predicting these critical properties.

On the contrary, unlike the mentioned parameters, the prediction of magnesium oxide (MgO), where the minimum R^2 values were 0.88996 with Gradient Boosting and 0.92485 with Random Forest Regression, has some challenges. These values are still acceptable, but they indicate room for improvement in the modeling approach for this component. ML results highlight the strength of Random Forest Regression and Gradient Boosting in accurately predicting the properties of biomass and coal blends in LIBS system. The results also point to exciting potential in applying machine learning and LIBS systems to deepen insights and enhance the use of these materials.

Acknowledgement

This material is based upon work supported by the U.S. Department of Energy, Office of Science, Office of Fossil Energy and Carbon Management under Award Number DE-C0022696.

Methane Conversion to Methanol: Technological Advances and Process Optimization

Omar M. Basha [a,*], Rui Wang [b,c], Isaac Gamwo [b], and Badie I. Morsi [b,c]

[a] ONB Engineering Research and Technical Services LLC, Delaware, USA

[b] U.S. Department of Energy, National Energy Technology Laboratory, P.O. Box 10940, Pittsburgh, PA 15236, USA

[c] Department of Chemical and Petroleum Engineering, University of Pittsburgh Pittsburgh, PA 15261, USA

Introduction to Methane and Methanol

Methane as a Resource and Greenhouse Gas

- Methane (CH_4) is the primary component of natural gas and a significant energy resource.
- As a greenhouse gas, methane is over 25 times more potent than CO_2 in trapping heat in the atmosphere.
- Mitigating methane emissions can play a crucial role in climate change mitigation efforts.

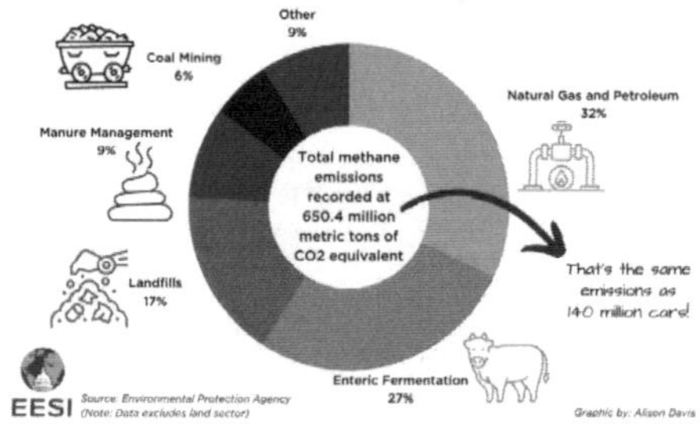

Methanol: A Versatile Platform Chemical

- Methanol (CH_3OH) is a key platform chemical with diverse applications:
  - Fuel Synthesis: Methanol can be used directly as a fuel or converted into synthetic fuels.
  - Chemical Feedstock: It is a precursor for producing formaldehyde, acetic acid, and other chemicals.
  - Energy Carrier: Methanol is seen as a potential energy carrier in the shift to renewable energy systems.

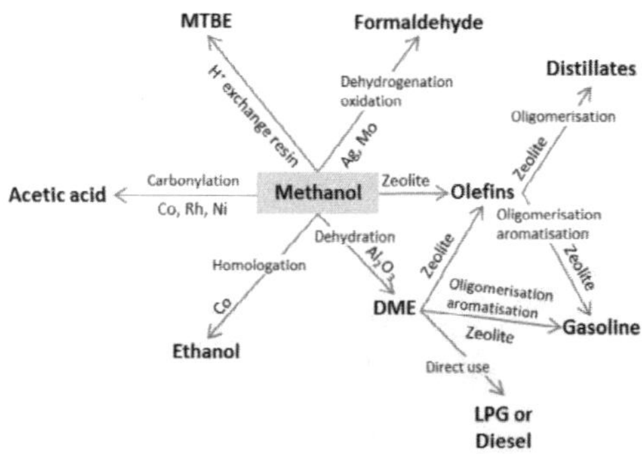

2

Overview Methane Conversion Methods

One-Step Direct Methane-to-Methanol (DMTM) Process

•**Main Reaction**:
Methane (CH_4) → Methanol (CH_3OH)

•**Catalyst Involved**: Include catalysts such as copper-based or iron-based.

•**Challenges**:
- Catalyst selectivity issues (may produce CO and CO_2 as by-products).
- Catalyst stability.
- Low yield of methanol due to by-products.

•**Advantages**:
- Simpler process (one-step reaction).
- Lower capital costs.

Two-Step Methane-to-Methanol Process

•**Step 1**: **Methane → Syngas** ($CH_4 + O_2/H_2O \rightarrow CO + H_2$)
- Steam reforming or partial oxidation.

•**Step 2**: **Syngas → Methanol** ($CO + 2H_2 \rightarrow CH_3OH$)

•**Catalyst Involved**: Common catalysts include copper-zinc oxides for methanol synthesis.

•**Advantages**:
- Better control over reaction parameters and product distribution.
- Proven technology with industrial scalability.

•**Challenges**:
- More complex and capital-intensive due to the syngas generation and separation steps.
- Higher operational costs.

3

Methane to Methanol Technologies

TRL 1-3: Concept, Early Stage Research (Exploratory)

Biological Methane Conversion
- **Key Points**: Uses methanotrophic bacteria; low conversion rates, scalability issues.
- **Challenges**: Difficult to scale up, limited to laboratory research.

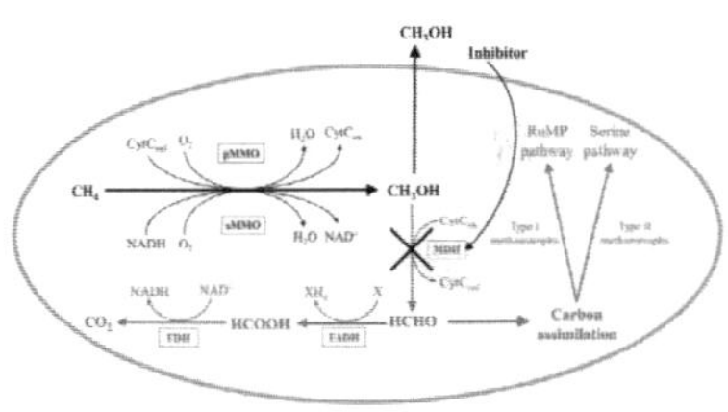

Photo-Catalysis for Methane to Methanol
- **Key Points:** Light-driven methane conversion using catalysts like TiO_2.
- **Challenges:** Low efficiency, catalyst stability, and poor light utilization.

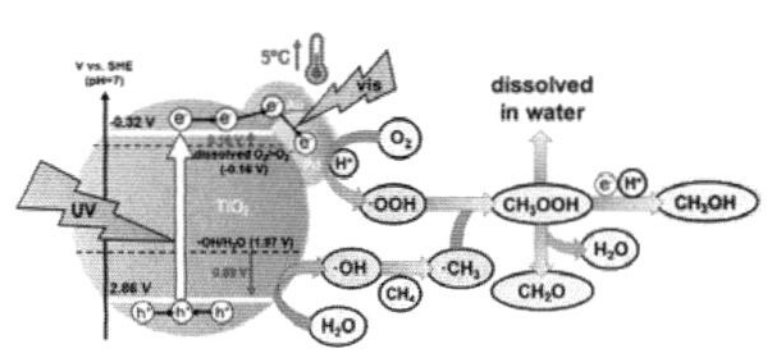

Oxidative Coupling of Methane (OCM)
- **Key Points:** Converts methane into higher hydrocarbons, leading to methanol.
- **Challenges:** High temperature, difficult to control product distribution.

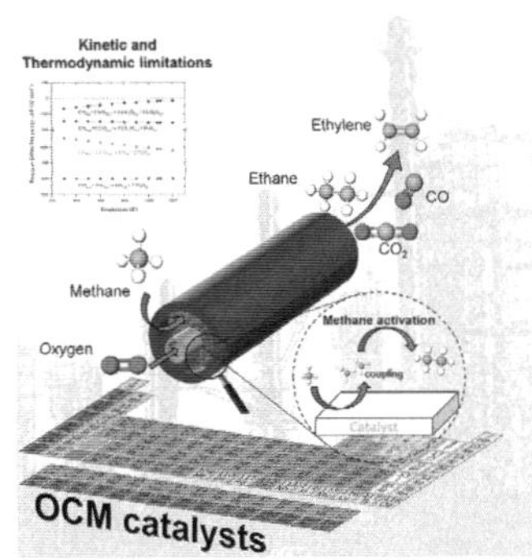

Cai, Xiaojiao, Siyuan Fang, and Yun Hang Hu. "Unprecedentedly high efficiency for photocatalytic conversion of methane to methanol over Au–Pd/TiO 2–what is the role of each component in the system?." *Journal of Materials Chemistry A* 9.17 (2021): 10796-10802.
Bjorck, Charlotte E., Paul D. Dobson, and Jagroop Pandhal. "Biotechnological conversion of methane to methanol: evaluation of progress and potential." *AIMS Bioengineering* 5.1 (2018): 1-38.
Ortiz-Bravo, C. A., Chagas, C. A., & Toniolo, F. S. (2021). Oxidative coupling of methane (OCM): An overview of the challenges and opportunities for developing new technologies. *Journal of Natural Gas Science and Engineering*, 96, 104254.

4

Methane to Methanol Technologies

TRL 4-5: Lab Validation and Early Prototype

Membrane-Assisted Methane Conversion

- **Key Points**: Combines membranes with methane conversion, aiming for efficient separation.
- **Challenges**: Membrane fouling, durability, and high costs.

Supercritical Water Oxidation (SCWO) for Methane

- **Key Points**: Uses high-pressure, high-temperature water conditions for methane oxidation.
- **Challenges**: Reactor complexity and high operational costs.

Non-Thermal Plasma-Assisted Conversion

- **Key Points**: Uses plasma for methane activation, enabling low-temperature conversion.
- **Challenges**: Energy inefficiency and scaling issues.

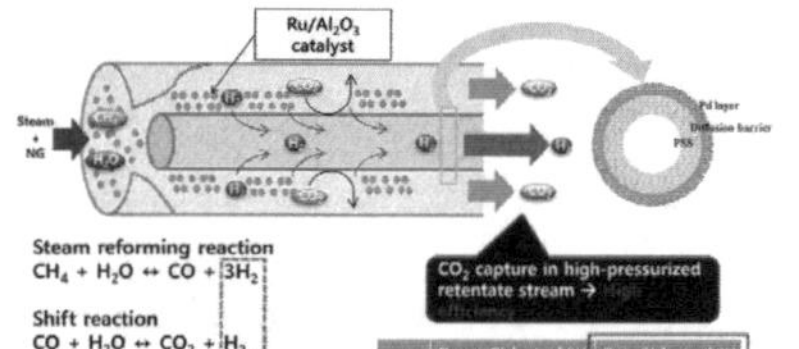

	Perm. Side (vol.%)	Ret. Side (vol.%)
H_2	97.84	3.5
CO	0.95	6.5
CH_4	0.65	22.2
CO_2	0.56	67.8

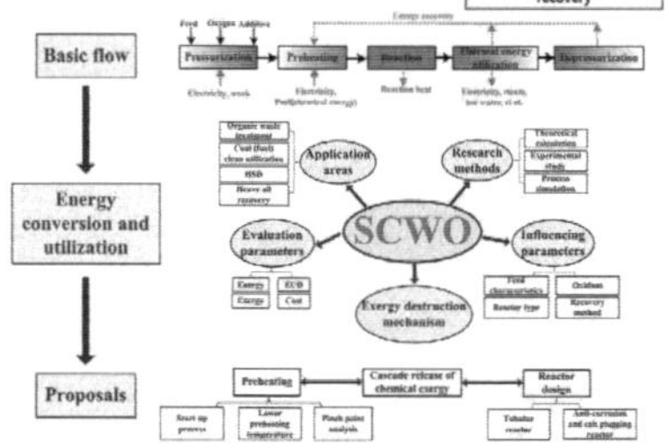

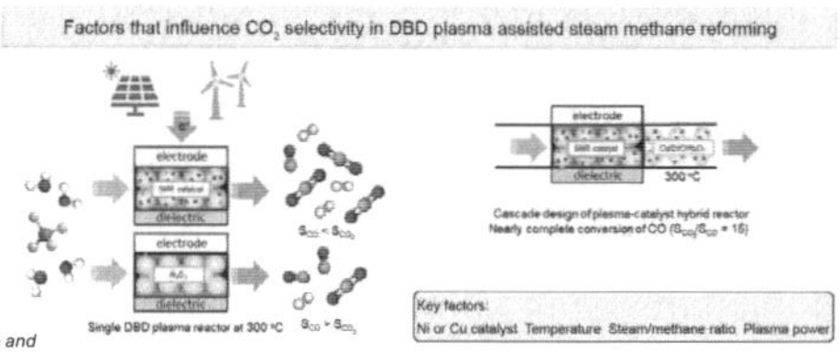

Geng, F., Haribal, V. P., & Hicks, J. C. (2022). Non-thermal plasma-assisted steam methane reforming for electrically-driven hydrogen production. *Applied Catalysis A: General*, *647*, 118903.
Zhang, F., Li, Y., Liang, Z., & Wu, T. (2022). Energy conversion and utilization in supercritical water oxidation systems: A review. *Biomass and Bioenergy*, *156*, 106322.
Ortiz-Bravo, C. A., Chagas, C. A., & Toniolo, F. S. (2021). Oxidative coupling of methane (OCM): An overview of the challenges and opportunities for developing new technologies. *Journal of Natural Gas Science and Engineering*, *96*, 104254.

Methane to Methanol Technologies

TRL 6-7: Pilot and Demonstration

Membrane-Assisted Technologies (advanced Stage)

- **Key Points**: Combines methane conversion with selective membranes to improve separation efficiency. Key pilot plant testing is ongoing, and the technology has progressed beyond laboratory research.
- **Challenges:** Membrane fouling and high operational costs.
- **Current Deployment:** Undergoing pilot plant trials.

Advanced Catalysts for One-Step Methane to Methanol

- **Key Points**: Advanced catalytic materials (e.g., single-site catalysts) are being tested in pilot-scale reactors to improve methanol yield and selectivity.
- **Challenges:** Catalyst deactivation and large-scale process integration.
- **Current Deployment:** Pilots are underway, but full-scale industrial application still needs further testing.

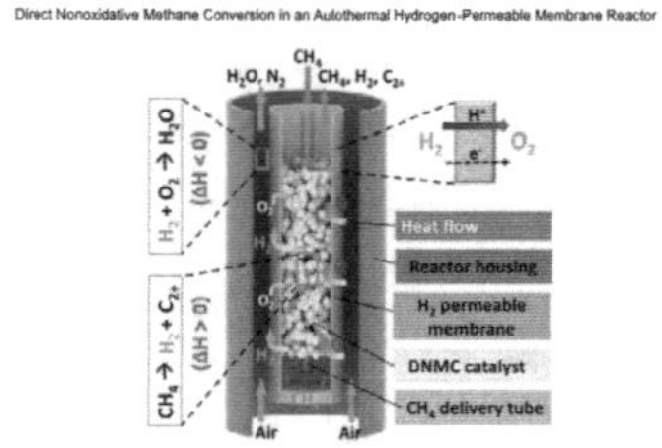

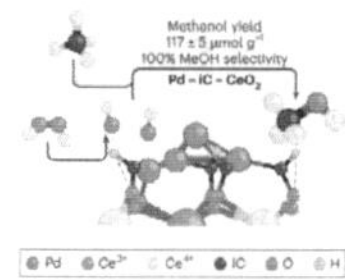

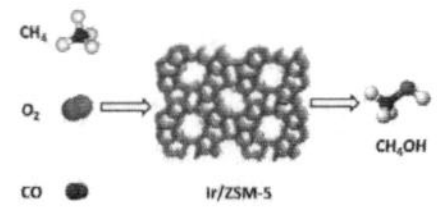

Sakbodin, M., Schulman, E., Cheng, S., Huang, Y. L., Pan, Y., Albertus, P., ... & Liu, D. (2021). Direct Nonoxidative Methane Conversion in an Autothermal Hydrogen-Permeable Membrane Reactor. *Advanced Energy Materials*, *11*(46), 2102782.
Li, Mengwei, Junjun Shan, Georgios Giannakakis, Mengyao Ouyang, Sufeng Cao, Sungsik Lee, Lawrence F. Allard, and Maria Flytzani-Stephanopoulos. "Single-step selective oxidation of methane to methanol in the aqueous phase on iridium-based catalysts." *Applied Catalysis B: Environmental* 292 (2021): 120124.
Groves, Alexandra R. "Methane to methanol." *Nature Synthesis* (2024): 1-1.

Methane to Methanol Technologies

TRL 8-9: Fully Commercialized and de-risked

Methane to Syngas to Methanol (Two-Step Process)

Syngas Generation Methods:

- Steam Methane Reforming (SMR): Methane reacts with steam to produce syngas ($CH_4 + H_2O \rightarrow CO + 3H_2$).
- Autothermal Reforming (ATR): A combination of partial oxidation and steam reforming, balancing oxygen and steam inputs to generate syngas efficiently.
- Partial Oxidation (POX): Methane is partially oxidized with oxygen to form syngas ($CH_4 + 0.5\ O_2 \rightarrow CO + 2H_2$), often used as part of the syngas generation in conjunction with SMR or ATR.

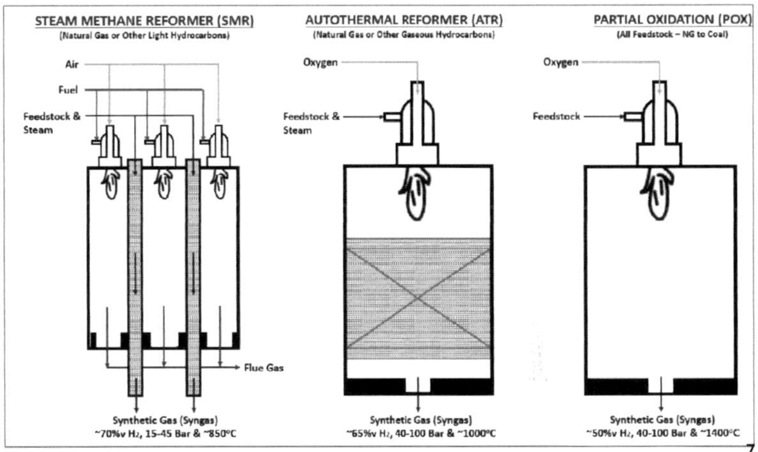

Methanol Synthesis:

- Once syngas is generated, it is converted to methanol using a copper-based catalyst in a separate reactor ($CO + 2H_2 \rightarrow CH_3OH$).

7

Challenges in DMTM

Methane to methanol (DMTM) lacks an industrially viable process due to high barriers in methane activation and product stability.

Key Challenges

- **Methane Activation**: Transition metals struggle with activation due to methane's symmetry and bond strength.
- **Product Reactivity**: Methanol's bonds are weaker than methane's, making it prone to further oxidation into unwanted byproducts like CO_2 and CO.

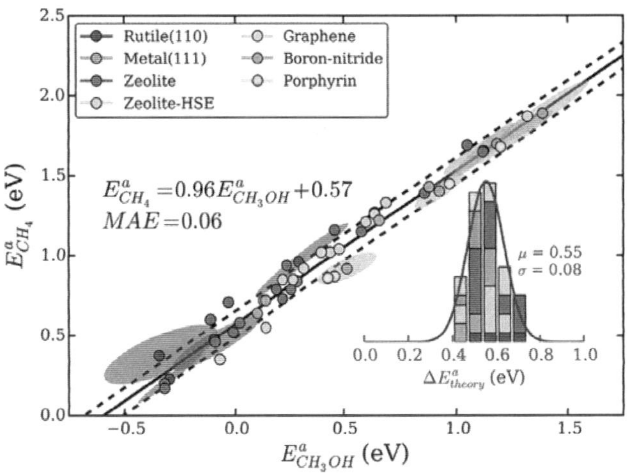

Latimer, A. A., Kakekhani, A., Kulkarni, A. R., & Nørskov, J. K. (2018). Direct methane to methanol: the selectivity–conversion limit and design strategies. *ACS catalysis*, *8*(8), 6894-6907.

8

Understanding Selectivity–Conversion Trade-Off

- In methane to methanol conversion, there is an inherent challenge between achieving high selectivity for methanol while maintaining meaningful methane conversion rates. This trade-off is primarily driven by temperature constraints and the differing activation energies required to convert methane and methanol.

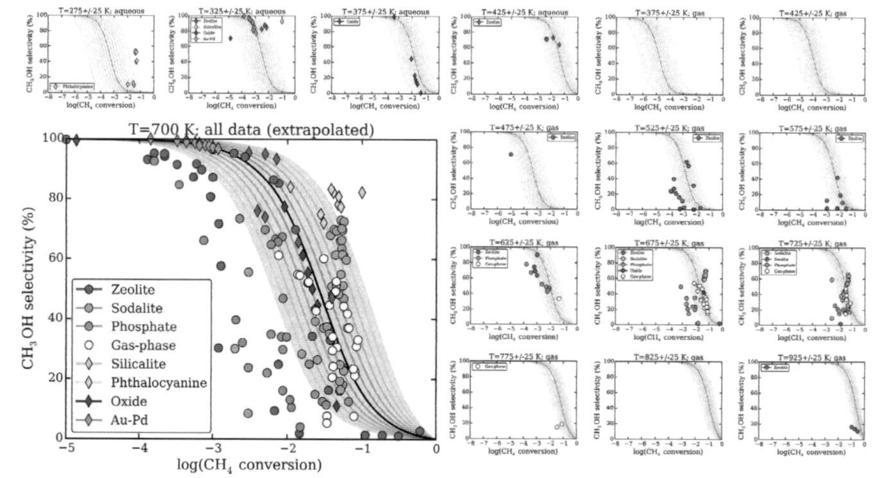

Latimer, A. A., Kakekhani, A., Kulkarni, A. R., & Nørskov, J. K. (2018). Direct methane to methanol: the selectivity–conversion limit and design strategies. *ACS catalysis*, *8*(8), 6894-6907.

9

Catalyst Design and Innovations in One-Step Processes

Innovations in Catalyst Design

- **Non-transition Metals**: Some non-transition metals show promise in activating methane at low temperatures (DFT-predicted barriers).
- **Zeolite-Based Protecting Groups**: Zeolites with metals like Fe and Cu can achieve selective oxidation to methanol with improved yields (up to 98% selectivity).
    - Stepwise processes using zeolites offer higher methanol selectivity but are energy-intensive and rely on expensive oxidizing agents.
- **Collector Concept**: Introduces a "methanol collector" to remove methanol during reactions, preventing further oxidation. This helps bypass the selectivity–conversion trade-off.

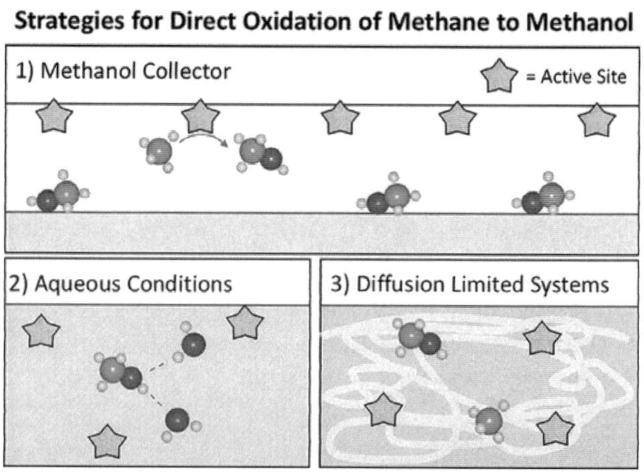

Latimer, A. A., Kakekhani, A., Kulkarni, A. R., & Nørskov, J. K. (2018). Direct methane to methanol: the selectivity–conversion limit and design strategies. *ACS catalysis*, *8*(8), 6894-6907.

10

Reactor Design and Innovations in One-Step Processes

- **Control and Optimization of Reaction Conditions:** Reactor design helps maintain optimal temperature and pressure to address the selectivity-conversion trade-off.
- **Innovative Configurations:** Microreactors, membrane reactors, and helical reactors enhance methane-catalyst contact, improving selectivity and reducing by-products.
- **Multi-Stage Reactors:** Separating methane activation from methanol production boosts control and improves selectivity.
- **Catalyst Integration:** Advanced catalyst beds or fluidized systems improve conversion rates and catalyst stability.

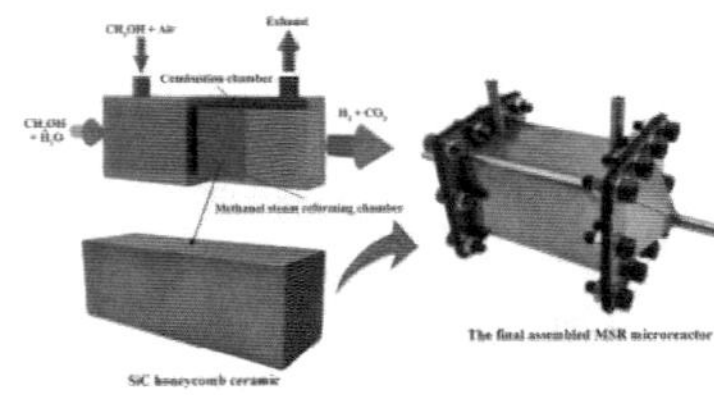

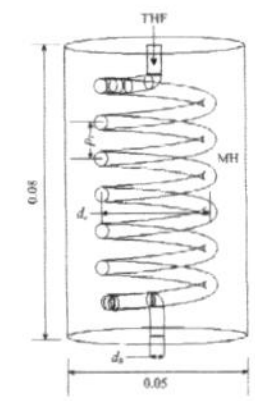

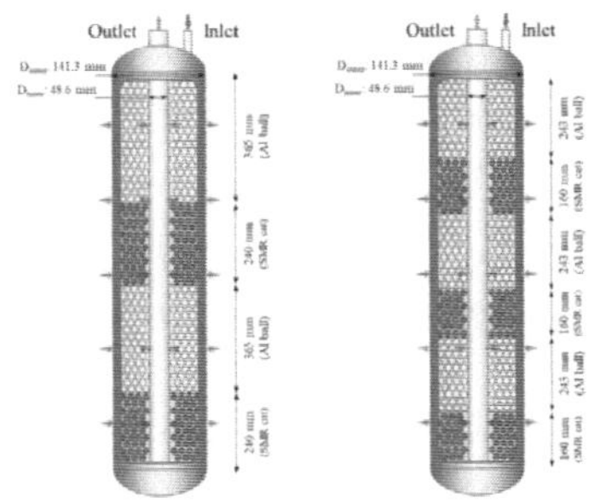

(a) Two-staged reactor (b) Three-staged reactor

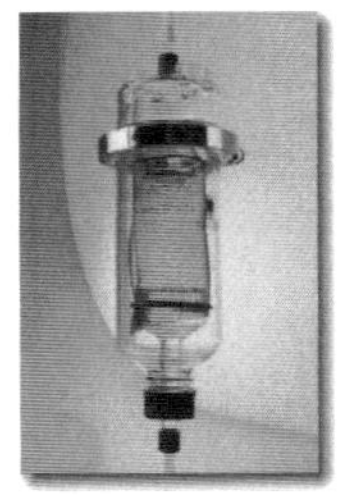

Comparative Assessment

Aspect	One-Step Process	Two-Step Process
Capital Investment	Lower capital cost due to simpler process design	High capital cost due to syngas production (60% of total)
Methane Conversion Rate	Lower (5-10% methane conversion)	Higher (20-30% methane conversion)
Operational Complexity	Simple process but challenging to scale effectively	Complex but established with proven product control
Product Yield	Small quantities of methanol, requires recycling	Larger quantities with high methanol selectivity
Scalability	Difficult to scale while maintaining efficiency	Highly scalable for large-scale production
Process Efficiency	Lower efficiency due to difficulty in maintaining selectivity	Higher efficiency with better carbon and energy recovery
Industry Maturity	Still under development with limited applications	Established with wide-scale industrial use
Key Trade-Offs	Cheaper upfront but less efficient and harder to scale	Higher cost but more efficient and suitable for large-scale applications

de Klerk, A. (2015). Engineering evaluation of direct methane to methanol conversion. *Energy Science & Engineering*, *3*(1), 60-70.

Comparative Assessment

Aspect	One-Step Process	Two-Step Process
Capital Cost (per ton of methanol)	\$300–\$500 million	\$700–\$1,200 million
Operating Cost (per ton of methanol)	\$120–\$160 per ton	\$80–\$100 per ton
Methanol Yield (per ton of methane)	50–60%	85–95%
CO_2 Emissions (per ton of methanol)	1.2 tons of CO_2	0.6 tons of CO_2
Energy Efficiency (%)	25–30%	50–55%
Technology Readiness Level (TRL)	4–6	7–9
Global Adoption Rate	Less than 10 pilot plants	95% of global methanol production
Carbon Efficiency (%)	35–40%	65–68%

De Klerk, A. (2019). "Engineering Evaluation of Direct Methane to Methanol Conversion Process for Small-Scale GTL Applications." Industrial & Engineering Chemistry Research, 58(24), 10468–10481. doi:10.1021/acs.iecr.9b01459.
Methanol Institute. (2020). "Methanol as a Commodity: Production Processes and Applications." Methanol.org.
Gesser, H. D., Hunter, N. R.. (1998). "Direct Conversion of Methane to Methanol: From Chemistry to Catalysis." Journal of Natural Gas Science and Engineering, 15(1), 27–38. doi:10.1016/j.jngse.2013.01.013.
Kuo, J. C.. (2014). "A Review on Direct Methane Conversion to Methanol Technologies." Applied Catalysis B: Environmental, 152(4), 191–204. doi:10.1016/j.apcatb.2013.10.008.
Arutyunov, V. S.. (2015). "Theoretical Perspectives and Real-Life Challenges of Methane to Methanol Conversion." Catalysis Today, 251(1), 69–81. doi:10.1016/j.cattod.2014.11.029.

13

Future Outlook and Research Directions

- **Catalyst Selectivity Improvement:** Ongoing research aims to enhance catalyst selectivity in DMTM processes to reduce by-products and improve efficiency.
- **Hybrid Process Development:** New hybrid approaches are combining the simplicity of one-step methods with the scalability of two-step processes.
- **Integration with CCU Technologies:** Future integration of DMTM with carbon capture technologies can reduce CO_2 emissions and produce methanol sustainably.
- **Innovative Reactor Designs:** Novel designs like helical reactors improve heat and mass transfer, optimizing reaction conditions and selectivity.

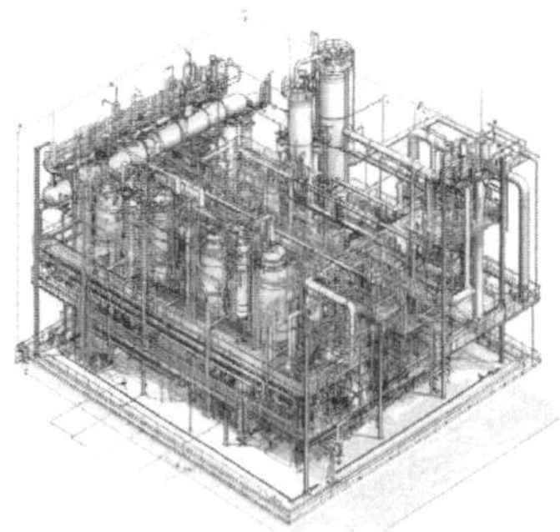

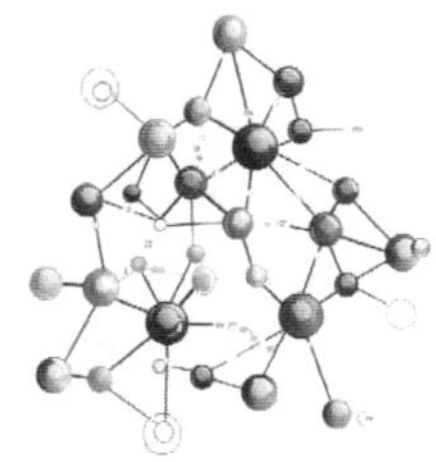

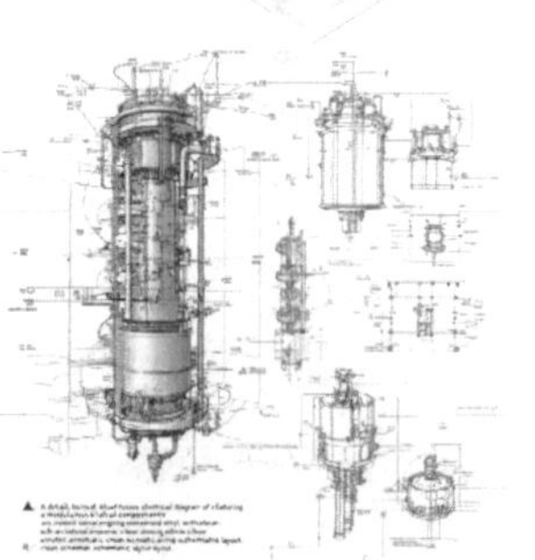

14

Concluding Remarks

- Methane-to-methanol conversion offers significant potential benefits but faces technical and economic challenges.
- One-step processes promise simplicity but struggle with catalyst selectivity and scalability.
- Two-step processes are more established, offering better control but requiring higher capital investments.
- Future research will focus on optimizing catalysts, novel reactor designs like helical reactors, and integration with carbon capture and utilization technologies.

15

TEA of the SMR-CCS Process for Blue H_2 Production Using Thirty-Six Physical Solvents

Rui Wang, Husain E. Ashkanani, Nicholas S. Siefert, Robert L. Thompson, Jeffrey Culp, Kathryn Smith, Janice A. Steckel, Wei Shi, Isaac K. Gamwo, David Hopkinson, Victor Kasuma, and Badie I. Morsi

Oct 8, 2024

U.S. DEPARTMENT OF ENERGY

University of Pittsburgh

NATIONAL ENERGY TECHNOLOGY LABORATORY

Authors

Rui Wang	US-DOE/NETL, PITT
Husain E. Ashkanani	US-DOE/NETL, KUWAIT UNIV.
Nicholas S. Siefert	US-DOE/NETL
Robert L. Thompson	US-DOE/NETL
Jeffrey Culp	US-DOE/NETL
Kathryn Smith	US-DOE/NETL
Janice A. Steckel	US-DOE/NETL
Wei Shi	US-DOE/NETL
Isaac K. Gamwo	US-DOE/NETL
David Hopkinson	US-DOE/NETL
Victor Kasuma	LRST/NETL
Badie I. Morsi	US-DOE/NETL, PITT

Outline

- Introduction and background
- Research objectives
- Results and discussion
- Concluding remarks

Global warming and CO_2 emission

- 2023 Global GHG emissions was 58.79 GT
- 60% of emission comes from energy systems and industrial sector.
- CO_2 emissions have detrimental impacts on humans and the environment.

Global 2023 GHG emissions in CO_2 equivalent [1]

Sector	Total (GT)	Share
Energy Systems	21.27	36%
Industrial sector	14.40	24%
Agriculture	11.90	20%
Transport	7.98	14%
Buildings	3.24	6%
Total	58.79	100%

(1 Gigaton = 10^9 ton)

1. Worldemissionclock. (2024). GHG emissions by sectors in 2024. Available: https://worldemissions.io/.

Impacts of CO_2 emissions on humans and the environment

- Increasing frequency and intensity of extreme heatwaves, droughts, hurricanes and floods.
- Rising sea levels, resulting in inundation of low-lying areas, and displacement of communities.
- Increasing potential conflicts among nations arising from resource scarcity and human migration.
- Deteriorating human health, such as increased rates of respiratory and heat-related illness, and the spread of disease-carrying insects.

2. P. M. Kellstedt, S. Zahran, and A. Vedlitz, "Personal efficacy, the information environment, and attitudes toward global warming and climate change in the United States," Risk Analysis: An International Journal, vol. 28, no. 1, pp. 113-126, 2008.

Technologies for CO_2 capture from industrial and energy systems

- CO_2 capture from pre-combustion applications.
- CO_2 capture from oxy-fuel combustion technologies.
- CO_2 capture from post-combustion applications.
- CO_2 capture from the Steam Methane Reforming process with carbon capture and sequestration (SMR-CCS).
- OUR RESEARCH WAS FOCUSING ON THE LAST TECHNOLOGIES

3. E. S. Rubin, C. Chen, and A. B. Rao, "Cost and performance of fossil fuel power plants with CO2 capture and storage," Energy policy, vol. 35, no. 9, pp. 4444-4454, 2007.

Hydrogen as a clean fuel

- H_2 is a clean fuel as water (H_2O) is the only biproduct.
- Can be converted to ammonia.
- Used in oil refineries.
- H_2 fuel cells are used in vehicles, and numerous industrial and residential applications

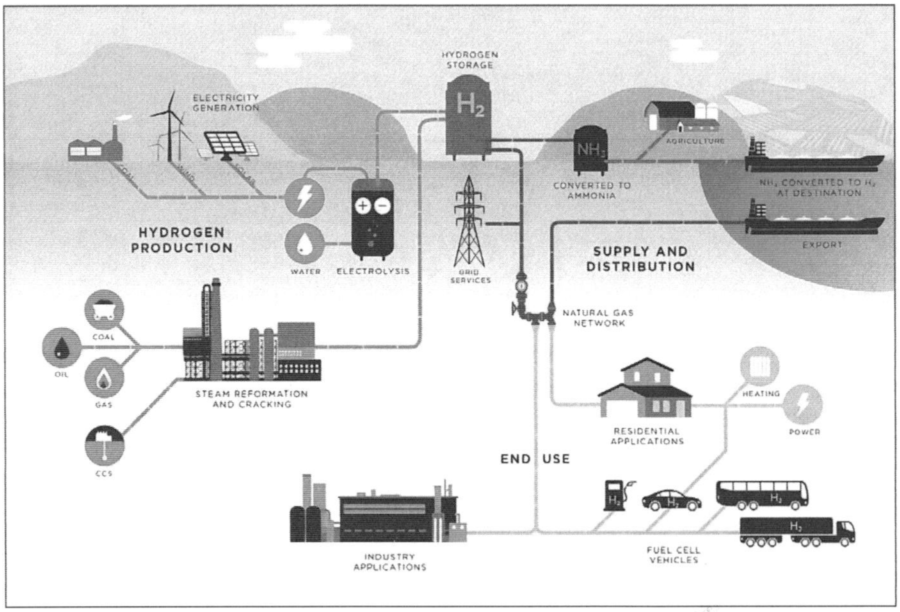

4. Vitalie Ciobanu, "Hydrogen Economy." https://www.linkedin.com/pulse/hydrogen-economy-20-24022023-vitalie-ciobanu/

Hydrogen production

- Gasification of coal and/or biomass, methane pyrolysis, steam methane reforming (SMR).
- Water Electrolysis.
- SMR contributes 95 % H_2 production in the US.
- SMR-CCS process is to produce blue H_2.

$$CH_4 + H_2O \rightleftharpoons CO + 3H_2$$

$$C_2H_6 + 2H_2O \rightleftharpoons 2CO + 5H_2$$

$$C_3H_8 + 3H_2O \rightleftharpoons 3CO + 7H_2$$

$$C_4H_{10} + 4H_2O \rightleftharpoons 4CO + 9H_2$$

$$CO + H_2O \rightleftharpoons CO_2 + H_2$$

5. Haryanto, Agus, Sandun Fernando, Naveen Murali, and Sushil Adhikari. "Current status of hydrogen production techniques by steam reforming of ethanol: a review." Energy & Fuels 19, no. 5 (2005): 2098-2106.

DOE/NETL SMR-CCS blue hydrogen (2018)

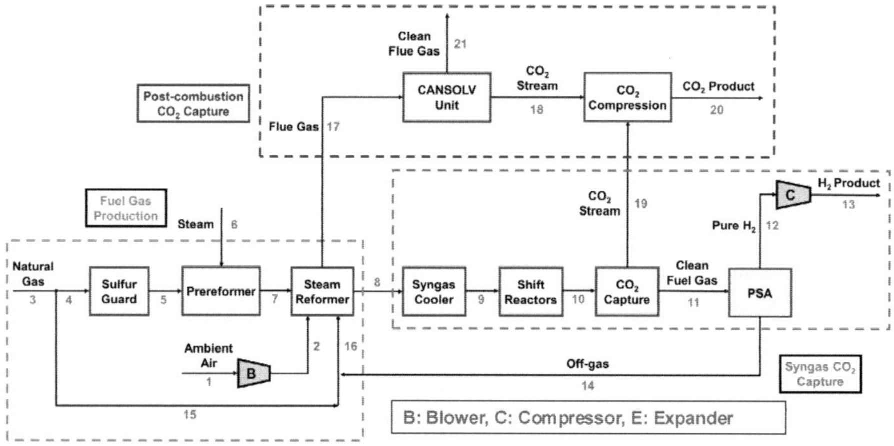

- **MDEA is used for CO_2 capture from the shifted fuel gas (stream #10).**
- **CANSOLV is used for post-combustion CO_2 capture (stream # 17).**
- **Post-combustion CO_2 capture represents 35% of the process CAPEX [6].**

6. E. Lewis et al., "Comparison of commercial, state-of-the-art, fossil-based hydrogen production technologies," National Energy Technology Laboratory (NETL), Pittsburgh, PA, Morgantown, WV, 2022.

Research objectives

- **To modify the 2018 DOE/NETL's SMR-CCS process to improve its performance and flexibility**
- **To model the modified process in Aspen Plus V.12.1 and conduct TEA to assess its economic feasibility**
- **To draw a direct comparison with the 2018 DOE/NETL's process.**

Modified SMR-CCS process flow diagram developed in Aspen Plus V.12.1

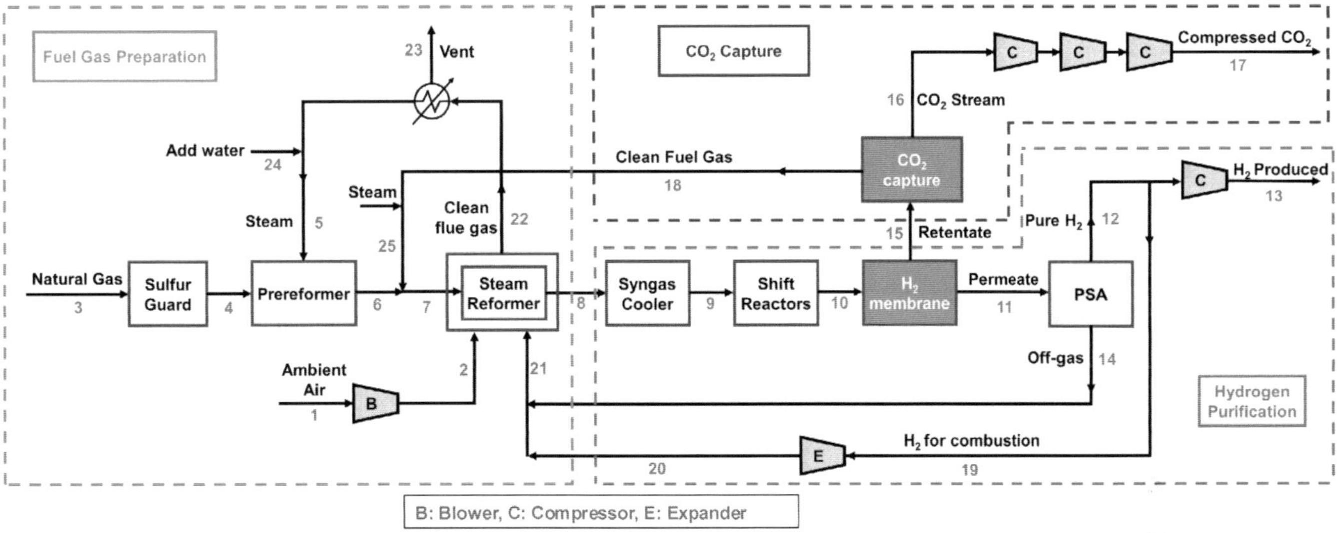

Key modifications

1) In the 2018 process, the entire shifted fuel gas coming from the WGS reactor was sent directly to the CO_2 capture process using methyl diethanolamine (MDEA) as a chemical solvent. In the modified process, the entire shifted fuel gas was sent directly to a hydrogen-selective membrane (polybenzimidazole - PBI) to separate it into an H_2-rich "permeate" and a CO_2-rich "retentate" .

2) In the 2018 process, the clean fuel gas from the CO_2 capture process was sent to a pressure swing adsorption (PSA) unit. In the modified process, only the permeate was sent to a PSA unit operating at adsorption pressure of 24.5 bar and desorption pressure of 1 bar to produce a pure H_2 stream and an off-gas containing (CH_4, CO, and H_2).

3) In the 2018 process, the heat required for the SMR reactor was provided by combusting an ambient air stream with a gas mixture consisting of the PSA off-gas and a split stream of the natural gas (CH_4) feed. In the modified process, the heat required for the SMR reactor was provided by combusting an ambient air stream (O_2) with a gas mixture consisting of the PSA off-gas and a split stream of the pure H_2 from the PSA.

Key modifications (cont.)

4) In the 2018 process, the resulting combustion flue gas stream contained about 7.5 mol% CO_2, which was captured in the CANSOLV unit containing concentrated aqueous amines. In the modified process, the resulting post-combustion flue gas contained negligible amount of CO_2 (0.9 mol%), leading to the elimination of the expensive CANSOLV unit. The water vapor in this flue gas was condensed and recycled back to the pre-reformer.

5) In the 2018 process, the entire fuel gas was sent to the CO_2 capture process using a chemical solvent (MDEA) and the clean gas was used as a fuel. In the modified process, only the retentate was sent to the CO_2 capture process, wherein with the aid of a compressor, the partial pressure of CO_2 was increased allowing the use of several effective physical solvents for CO_2 capture. The clean fuel gas, containing methane (CH_4), was recycled back to the SMR.

Key modifications (cont.)

6) In the 2018 process, the entire fuel gas was sent to the CO_2 capture process using a chemical solvent (MDEA) and the clean fuel gas was then purified via PSA to produce pure H_2, while the off-gas was used as a fuel for the SMR. In the modified process, only the retentate was sent to the CO_2 capture process, wherein with the aid of a compressor, the partial pressure of CO_2 was increased allowing the use of several effective physical solvents for CO_2 capture. The clean fuel gas, containing methane (CH_4), was recycled back to the SMR.

7) In the 2018 process, the CO_2 captured using MDEA and CANSOLV was dehydrated and compressed in preparation for subsequent sequestration; and the main product is 99.8% pure H_2. In the modified process, the CO_2 captured using 36 physical solvents was dehydrated and compressed in preparation for subsequent sequestration; and the main product was H_2 with 99.863% purity.

Physical comparison between the modified and the 2018 DOE/NETL SMR-CCS process

	Modified SMR-CCS Process	2018 DOE/NETL SMR-CCS Process
Energy source for heating the SMR	The reaction between O_2 in the ambient air with a combined gas stream made of the PSA off-gas and a split stream of pure H_2.	The reaction between O_2 in the ambient air with the PSA off-gas and a split stream of natural gas feed.
Flue gas treatment	The flue gas is almost CO_2-free (0.9 mol%).	The flue gas (7.5 mol% CO_2) is sent to the CANSOLV.
Shifted fuel gas treatment	The shifted fuel gas is separated using a polymeric H_2 membrane into a H_2-rich permeate and a CO_2-rich retentate.	The total shifted fuel gas is sent directly to the CO_2 capture absorber. (No Membrane)
CO_2 capture from shifted fuel gas	Thirty-six physical solvents were used for CO_2 capture from the retentate.	One chemical solvent (MDEA) was used for CO_2 capture.
CO_2 capture efficiency	97%	96%

Aspen Plus v.12.1 modeling of the SMR-CCS process for blue H_2 production

1. to develop a H_2-selective membrane in Aspen Custom Modeler V.12.1
2. to develop a pressure swing adsorption (PSA) in Aspen Adsorption V.12.1
3. to use 36 physical solvents (3 BACs, 11 ILs, 5 HCs, 7 OHCs, 2 NHCs, 5 CycHCs, and 3 Polymers) for CO_2 capture.
4. to calculate TEA values: CAPEX, OPEX, FEX, LCOC, and LCOH
5. to compared the CAPEX, OPEX, FEX, LCOC, LCOH, H_2/CO_2, and H_2/CH_4 of this modified process with those of the DOE/NETL.
6. to develop an artificial neural network (ANN) to predict the TEA results of the 36 solvents used in the development of the modified SMR-CCS process.

H_2 hollow fiber membrane in Aspen Custom modeler V.12.1

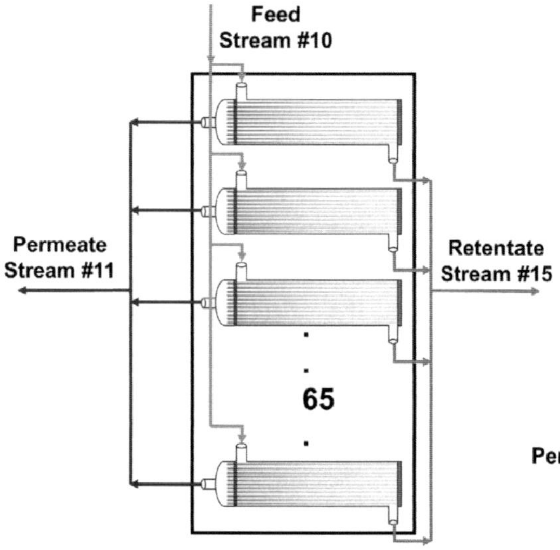

65 modules hollow fiber membrane

Shifted Fuel Gas	Stream # 10
Temperature, °C	204
Pressure, bar	24.5
Mole flow rate, kmol/h	36,689
Composition	mol%
H_2	62.82
H_2O	11.59
CO_2	10.69
CH_4	2.7
CO	0.3
N_2	11.9

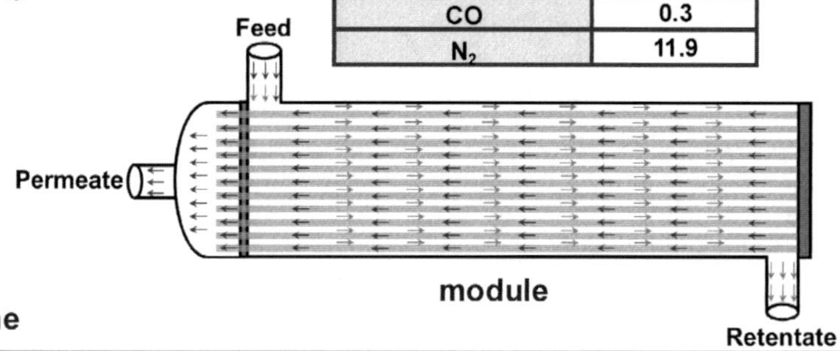

PBI hollow fiber H_2 membrane module

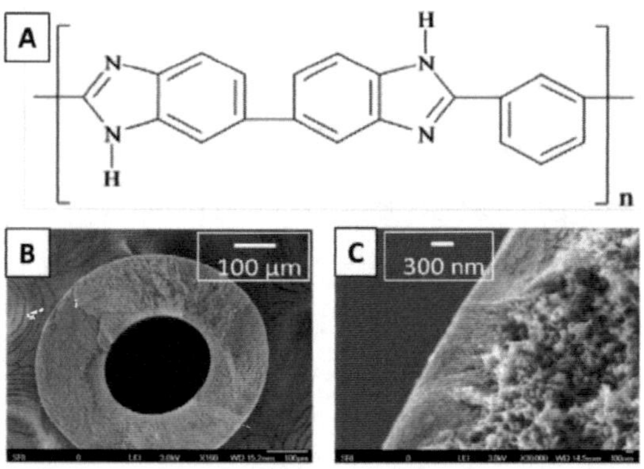

(A) Molecular structure of polybenzimidazole (PBI) (B) Structure of fiber (C) PBI membrane on the fiber outer surface [9]

GPC: Gas Permeance Coefficient

$$1\ \mathrm{GPC} = 7.5 \times 10^{-12}\ \frac{\mathrm{m^3(STP)}}{\mathrm{m^2 Pa \cdot s}} = 0.12\ \frac{\mathrm{mol}}{\mathrm{m^2 bar \cdot h}}$$

Table 1: Hollow fiber membrane specification [7]

Fiber inner diameter, m	2.5×10^{-4}
Fiber outer diameter, m	6.5×10^{-4}
Fiber length, m	0.8
Number of fibers in module	625,000
Module diameter, m	0.57

Table 2: PBI membrane permeance of gases [7,8]

Gases	GPC	Selectivity H_2/Gases
CH_4	0.919	234
CO	2.087	103
CO_2	5.375	40
H_2	215	1
H_2O	645	1/3
N_2	2.194	98

7. Siefert, N.S., Narburgh, S. and Chen, Y., 2016. Comprehensive exergy analysis of three IGCC power plant configurations with CO2 capture. Energies, 9(9), p.669.
8. E. Perea and I. Jayaweera, "Development and testing of a high-temperature PBI hollow-fiber membrane technology for pre-combustion CO2 capture," 2022.
9. Wales, M.D., Gebremichael, E., Wang, X., Perea, E., Jayaweera, P. and Jayaweera, I., 2021. Flue gas desulfurization (FGD) wastewater treatment using polybenzimidazole (PBI) hollow fiber (HF) membranes. Membranes, 11(6), p.430.

H_2 membrane results

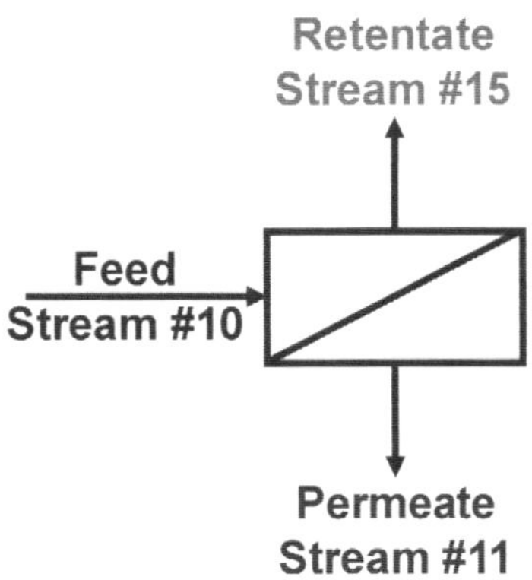

	Feed (stream #10)	Permeate (stream #11)	Retentate (stream #15)
Temperature, °C	204	204	204
Pressure, bar	24.5	5	23.6
Molar flow rate, kmol/h	36,689	19,254	17,435
		mol%	
H_2	62.82	78.46	45.54
H_2O	11.59	20.57	1.67
CO_2	10.69	0.64	21.8
CH_4	2.7	0.03	5.65
CO	0.3	0.01	0.63
N_2	11.9	0.3	24.71

2 PSA units with 4 adsorbers

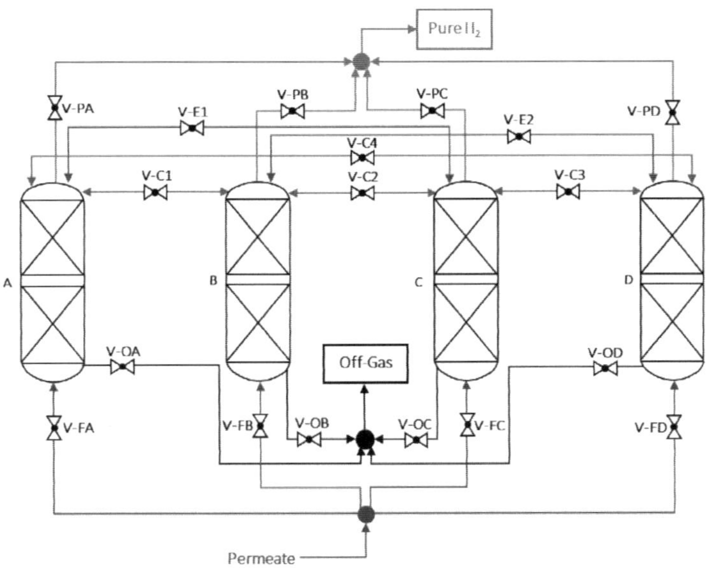

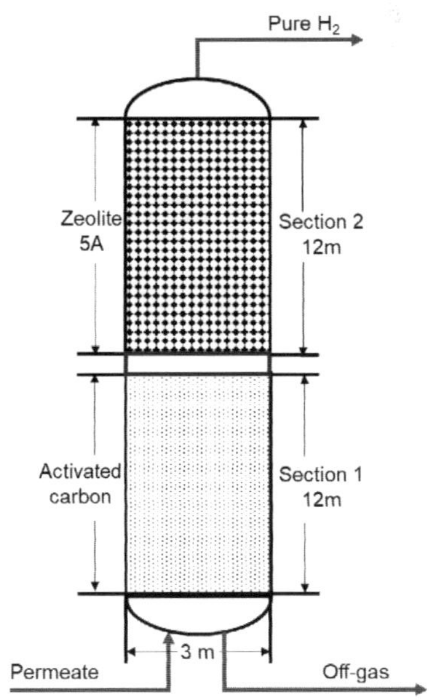

PSA dynamic modeling results

Bed stages

1. Adsorption of non-H_2 gases - H_2 production
2. Pressure equalization
3. Venting to off-gas
4. Purging non-H_2 gases with pure H_2

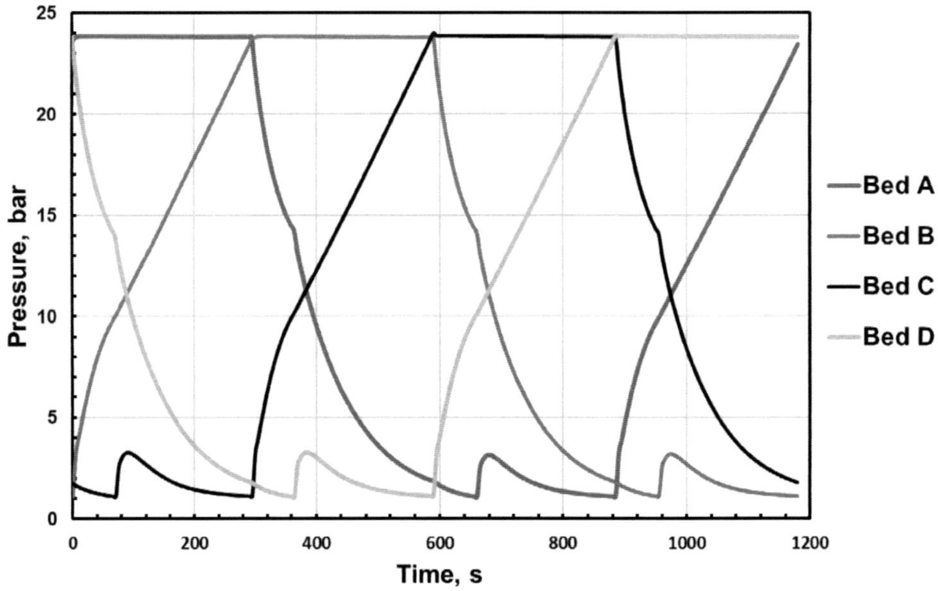

PSA adsorber and adsorbents

Total adsorber height, m	26
Lower section packing height, m	12
Inside radius (R_{Bi}), m	1.5
Outside radius (R_{Bo}), m	1.515
Heat capacity of the wall (C_{pw}), J/kg.K	502
Density of adsorber wall (ρ_w), kg/m³	7,830
Internal heat transfer coefficient (h_i), kJ/m³ K.s	3.85
External heat transfer coefficient (h_o), kJ/m³ K.s	1.42
Axial dispersion coefficient (D_L), m²/s	1.5×10^{-5}
Axial thermal conductivity coefficient (K_L), J/K·m·s	0.452

Adsorbent	Activated Carbon	Zeolite 5A
Type	Granular	Sphere
Normal pellet size, mesh	6-16	4-8
Average pellet size (R_P), mm	1.15	1.57
Pellet density (ρ_P), kg/m³	850	1,160
Adsorbent heat capacity, J/kg.K	1,046	921
Particle porosity, α	0.61	0.65
Bed density (ρ_B), kg/m³	482	746
Bed voidage, ϵ	0.433	0.357
Total void fraction, ϵ_t	0.779	0.775

10. J.-G. Jee, M.-B. Kim, and C.-H. Lee, "Adsorption characteristics of hydrogen mixtures in a layered bed: binary, ternary, and five-component mixtures," Industrial & engineering chemistry research, vol. 40, no. 3, pp. 868-878, 2001.
11. J. Yang, C.-H. Lee, and J.-W. Chang, "Separation of hydrogen mixtures by a two-bed pressure swing adsorption process using zeolite 5A," Industrial & engineering chemistry research, vol. 36, no. 7, pp. 2789-2798, 1997.

PSA results

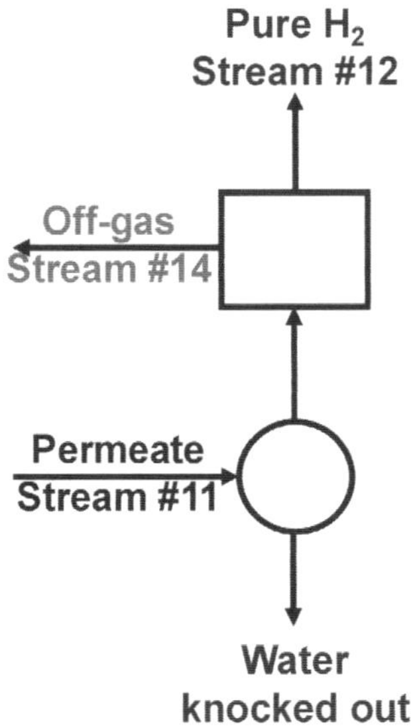

	Permeate (stream #11)	Pure H_2 (stream #12)	Off-gas (stream #14)	Water knocked out
Temperature, °C	204	12.21	25	3
Pressure, bar	5	23.86	1	5
Flow rate, kmol/h	19,254	12,542	2,752	3,960
	mol%			
H_2	78.46	99.983	93.27	0.00
CO_2	0.64	0.00	4.49	0.00
CH_4	0.03	0.00	0.20	0.00
H_2O	20.57	0.00	0.00	100
CO	0.01	0.00	0.05	0.00
N_2	0.30	0.02	1.99	0.00

99.983 mol% H_2 purity

CO_2 capture process in the SMR-CCS

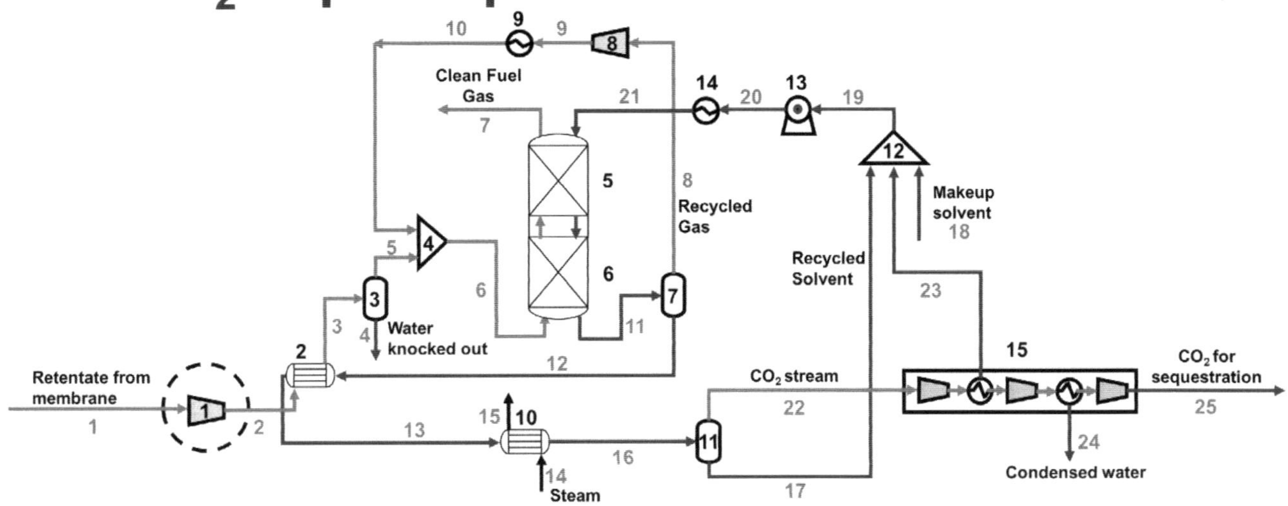

U-1.	Retentate gas compressor	U-6.	CO_2 absorber Packing 2	U-11.	Flash drum 2 (FD2)
U-2.	Cross-flow heat exchanger	U-7.	Flash drum 1 (FD1)	U-12.	Solvent makeup mixer
U-3.	Water knockout separator	U-8.	Recycled gas compressor	U-13.	Recycled solvent pump
U-4.	Gas mixer	U-9.	Recycled gas cooler	U-14.	Recycled solvent cooler
U-5.	CO_2 absorber Packing 1	U-10.	Solvent heater	U-15.	Multistage compressor with intercooling

Process hydraulics and mass transfer characteristics in packed-beds

- Packing used: Mellapak 250 Y (Structured packing):
- **Flooding:** Leva (1992) [7]
- **Two-phase pressure drop ($\Delta P/\Delta z$):** Billet and Schultes (1999) [8]
- **Liquid holdup (β_L):** Billet and Schultes (1999) [8]
- **Mass transfer coefficients (k_L and k_G):** Billet and Schultes (1999) [8]
- **Normalized specific wetted area (a_w/a):** Billet and Schultes (1999) [8]

12. Leva, M. (1992) 'Reconsider packed-tower pressure-drop correlations', Chemical engineering progress. American Institute of Chemical Engineers, 88(1), pp. 65–72.
13. R. Billet and M. Schultes, "Prediction of mass transfer columns with dumped and arranged packings: updated summary of the calculation method of Billet and Schultes," Chemical Engineering Research and Design, vol. 77, no. 6, pp. 498-504, 1999.

TEA calculations

$$\text{CAPEX}_{2023} = \sum_{equipment} (CAPEX_{2023,equipment})$$

$$\text{OPEX}_{2023} = C_{electricity} \sum W + C_{steam} + C_{O\&M}$$

$$\text{LCOC} = (f_{CR}/f_C)(CAPEX_{2023})/\dot{m}_{CO2} + OPEX_{2023}/\dot{m}_{CO2}$$

$$f_{CR} = \frac{i(1+i)^N}{(1+i)^N - 1}$$

W = total power required, MW
$C_{electricity}$ = Electricity cost, 2023-USD/MWh
C_{steam} = Cost of steam, 2023-USD/h
i = annual discount rate, %/year
N = project lifetime, years
f_{CR} = capital recovery factor, per year
$C_{O\&M}$ = Operating & Maintenance cost, 4% CAPEX
$\dot{m}_{CO2}$ = CO_2 captured, ton/h

Cost conversion and Installation Factor

- $CAPEX_{2023} = CAPEX_{year}(\frac{CEPCI_{2023}}{CEPCI_{year}})$

CEPCI = Chemical Engineering Plant Cost Index

- $Final(CAPEX_{2023}) = CAPEX_{2023}(IF)$

Unit	Installation Factor (IF)
Compressors	2.5
Heat exchangers	3.5
Pumps	4.0
Pressure vessels	4.0
Distillation columns	4.0

Year	CEPCI
2006	499.6
2017	567.5
2018	603.1
2019	607.5
2020	596.2
2021	708.0
2022	817.1
2023	797.9

14. T. R. Brown, "Cost Engineering: Equipment Purchase Cost," *Chemical Engineering*, pp. 51-53, 2019.
15. Towler, Gavin, and Ray K. Sinnott. *Chemical engineering design: principles, practice and economics of plant and process design.* Elsevier, 2008.
16. https://toweringskills.com/financial-analysis/cost-indices/

Process constraints and solvents used

- At least 97% CO_2 capture from the shifted fuel gas
- No flooding in the CO_2 absorber
- Packing height to internal diameter ratio (H/D) is ≥ 6
- Fuel gas ($CO + H_2 + CH_4$) in the CO_2 stream to sequestration is ≤ 0.5 mol%
- Water content in the CO_2 stream to sequestration is ≤ 600 ppm

Physical solvents used

- 3 Beta-Amino Carboxylate (BAC)
- 11 Ionic Liquids (ILs)
- 5 Hydrocarbons (HCs)
- 7 Oxygenated Hydrocarbons (OHCs)
- 2 Nitrogenized Hydrocarbons (NHCs)
- 5 Cyclical Hydrocarbons (CycHCs)
- 3 Polymers

CO_2 and H_2 solubilities in seven selected physical solvents

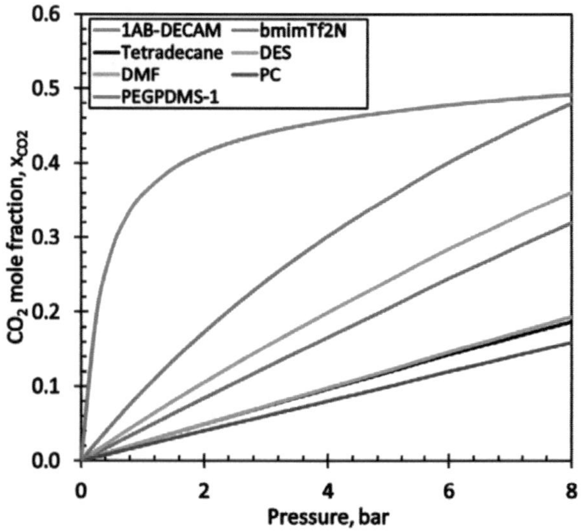

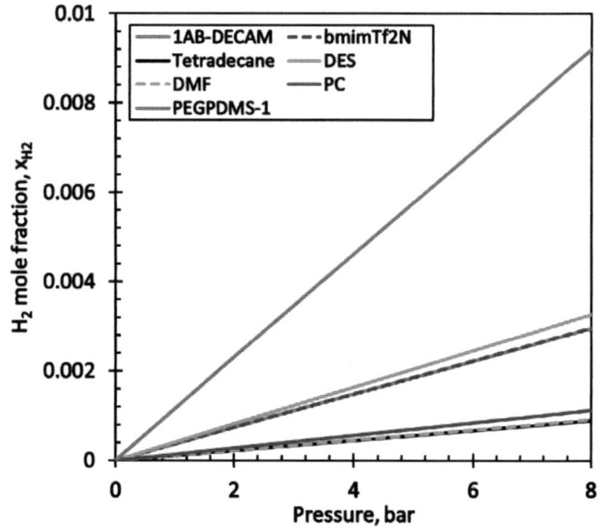

CO_2 capture process using 1AB-DECAM

1AB-DECAM	Retentate (stream #15)	Clean fuel gas (stream #7)	CO_2 stream (stream #15)
Temperature, °C	204	0	25
Pressure, bar	23.6	23.56	152.7
Molar flow rate, kmol/h	17,435	13,393	3,754
		(mol%)	
H_2	45.54	59.27	0.05
CO_2	21.80	0.50	99.45
CH_4	5.65	7.26	0.33
H_2O	1.67	0.01	0.08
CO	0.63	0.82	0.01
N_2	24.71	32.15	0.07
		(mass%)	
H_2	4.88	10.11	0.00
CO_2	50.98	1.84	99.79
CH_4	4.82	9.86	0.12
H_2O	1.6	0.02	0.03
CO	0.94	1.93	0.01
N_2	36.79	76.23	0.04

98.7% CO_2 capture

H_2O content 331 ppm

$CO + H_2 + CH_4 = 0.4\%$

Overall control volume of our SMR-CCS process

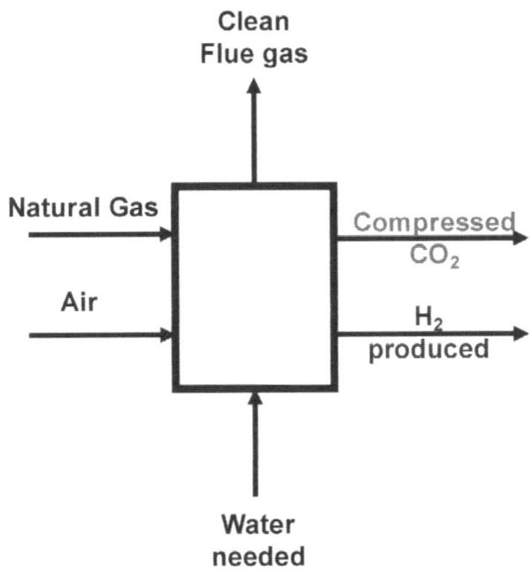

	Natural gas	Water needed	Air	Clean flue gas	Compressed CO_2	H_2 produced
Mass flow rate, ton/h	63.88	48.08	464.58	392.82	164.65	19.07
	(mass%)					
Ar	0.00	0.00	1.27	1.51	0.00	0.00
CH_4	87.00	0.00	0.00	0.00	0.12	0.00
C_2H_6	5.60	0.00	0.00	0.00	0.00	0.00
C_3H_8	1.80	0.00	0.00	0.00	0.00	0.00
C_4H_{10}	1.35	0.00	0.00	0.00	0.00	0.00
CO	1.63	0.00	0.00	0.00	0.01	0.00
CO_2	0.00	0.00	0.05	1.51	99.79	0.00
H_2	0.00	0.00	0.00	0.00	0.00	99.77
H_2O	0.00	100.00	0.62	3.78	0.03	0.00
N_2	2.61	0.00	75.06	89.17	0.04	0.23
O_2	0.00	0.00	23.00	4.03	0.00	0.00

TEA calculations of the SMR-CCS process

Parameter	Value
Cost of electricity	75.4 2023-$/MWh
Cost of PBI H_2 membrane	700 2023-$/m^2
i	2.01%/year
N	35 years
fc	0.9
f_{CR}	0.0447
$f_{O\&M}$	4% of CAPEX, $/year

$$\mathrm{LCOH} = \left(\frac{f_{CR}}{f_c}\right)\frac{\mathrm{CAPEX}_{2023}}{\dot{m}_{H2}} + \frac{\mathrm{OPEX}_{2023}}{\dot{m}_{H2}} + \frac{\mathrm{FEX}_{2023}}{\dot{m}_{H2}}$$

$$\mathrm{CAPEX\ to\ LCOH} = \left(\frac{f_{CR}}{f_c}\right)\frac{\mathrm{CAPEX}_{2023}}{\dot{m}_{H2}}$$

$$\mathrm{OPEX\ to\ LCOH} = \frac{\mathrm{OPEX}_{2023}}{\dot{m}_{H2}}$$

$$\mathrm{FEX\ to\ LCOH} = \frac{\mathrm{FEX}_{2023}}{\dot{m}_{H2}}$$

6. E. Lewis et al., "Comparison of commercial, state-of-the-art, fossil-based hydrogen production technologies," National Energy Technology Laboratory (NETL), Pittsburgh, PA, Morgantown, WV, 2022.

CAPEX of the SMR-CCS

CAPEX	Million 2023-$
Feedwater & Miscellaneous Equipment	112.99
Primary Reformer	86.97
Air Blower	5.07
Sulfur Guard Bed	1.96
Pre-reformer	4.76
Syngas Coolers	26.01
Water Gas Shift (WGS) Reactors	16.34
H_2 product compressor	4.66
H_2 product cooler	2.39
Accessory Electric Plant	12.54
Instrumentation & Control	21.82
Site Preparation, Improvements and Facilities	33.56
Buildings & Structures	20.76

CAPEX (cont.)	Million 2023-$
Water Separator	0.85
Compressor for PSA	8.39
Cooler for PSA	1.71
Pressure Swing Adsorber	14.23
H_2 product expander for combustion	2.80
Clean gas compressor	2.09
H_2 membrane	46.46
Absorber	4.23
Flash Drums	1.26
Rotating Equipment for CO_2 absorption	15.36
Heat Exchangers for CO_2 absorption	20.40
Initial Solvent Cost	6.82
Total	474.41

6. E. Lewis et al., "Comparison of commercial, state-of-the-art, fossil-based hydrogen production technologies," National Energy Technology Laboratory (NETL), Pittsburgh, PA, Morgantown, WV, 2022.

OPEX and FEX of the SMR-CCS

OPEX and FEX	Consumption Per Day	2023-$ Per Unit	M 2023-$/year
Water (gal/1000)	258.8	1.9	161.53
ZnO Sulfur Guard Catalyst (ft^3)	0.5	600	98.55
Pre-reformer Catalyst (ft^3)	0.6	1,250	246.38
Primary Reformer Catalyst (ft^3)	1.7	525	293.19
Water Gas Shift Catalyst (ft^3)	9.6	480	1,513.73
PSA Unit Adsorbent (ft^3)	4.5	150	221.74
CO_2 solvent makeup (Liter)	0.0	30	0.0
Electricity (MWh)	1,990.8	75.4	49,311
O&M			18,977
Natural Gas (MMBtu)	70,673	2.10	48,753

6. E. Lewis et al., "Comparison of commercial, state-of-the-art, fossil-based hydrogen production technologies," National Energy Technology Laboratory (NETL), Pittsburgh, PA, Morgantown, WV, 2022.

Performance of the SMR-CCS process (1AB-DECAM)

H_2 production rate, MM kg/year	150.03
CO_2 capture rate, MM kg/year	1,295
CAPEX-CO_2, MM 2023-$	48.91
OPEX-CO_2, MM 2023-$/year	27.50
LCOC, 2023-$/ton CO_2	20.80
CAPEX-H_2, MM 2023-$	474.41
OPEX-H_2, MM 2023-$/year	70.82
FEX-H_2, MM 2023-$/year	48.75
LCOH, 2023-$/kg H_2	0.93

Comparison with the DOE/NETL SMR-CCS process (2018)

	Our modified SMR-CCS Process (2018-USD)	DOE/NETL SMR-CCS Process (2018-USD)
H_2 production, ton/yr	150,031	157,002
CO_2 captured, ton/yr	1,295,349	1,516,900
CO_2 emitted to atmosphere, ton/yr	46,879	60,259
H_2/CO_2 molar ratio	67.85	57.32
H_2/CH_4 molar ratio	2.72	2.45
CO_2 capture efficiency, %	97	96
CAPEX, MM 2018-$	345.44	896.66
OPEX, MM 2018-$/yr	63.24	38.40
FEX, MM 2018-$/yr	102.61	130.36
LCOH, 2018-$/kg H_2	1.21	1.54

6. E. Lewis et al., "Comparison of commercial, state-of-the-art, fossil-based hydrogen production technologies," National Energy Technology Laboratory (NETL), Pittsburgh, PA, Morgantown, WV, 2022.

TEA results of the 36 Solvents

Solvent	H_2/CO_2 --	H_2/CH_4 --	CAPEX-CO_2 Million $	OPEX-CO_2 Million $/year	LCOC $/ton.$CO_2$	CAPEX-H_2 Million $	OPEX-H_2 Million $/year	LCOH $/kg$H_2$
EMA-NH-EMA	66.94	2.73	51.21	29.83	22.51	476.77	120.86	0.94
IBA-DECAM	64.17	2.73	51.37	29.65	22.42	477.05	121.96	0.95
1AB-DECAM	69.85	2.73	48.91	27.50	20.80	474.41	119.56	0.93
[aPy][Tf_2N]	65.98	2.63	80.81	71.30	52.08	504.45	158.82	1.25
[bmim][BF_4]	64.21	2.63	72.91	66.85	48.77	496.56	154.36	1.21
[bmim][$MeSO_4$]	59.58	2.64	108.00	141.57	101.92	531.66	228.25	1.73
[bmim][PF_6]	61.33	2.63	71.92	68.71	49.97	495.57	155.39	1.22
[bmim][Tf_2N]	60.95	2.63	54.59	42.72	31.41	480.04	134.31	1.07
[bmPyr][Tf_2N]	64.72	2.63	87.79	83.62	60.90	511.44	171.11	1.33
[emim][BF_4]	59.89	2.64	74.79	71.60	52.11	498.45	158.91	1.24
[emim][Tf_2N]	61.33	2.63	75.23	68.84	50.16	498.89	155.53	1.22
[hmim][$(C_2F_5)_3PF_3$]	59.17	2.64	62.74	49.98	36.72	487.21	140.26	1.11
[hmim][Tf_2N]	66.32	2.64	87.75	79.53	57.80	512.30	165.90	1.29
[omim][Tf_2N]	64.67	2.64	63.03	48.16	35.42	486.69	137.20	1.09
1-Heptene	66.69	2.64	108.92	153.71	109.84	532.57	240.44	1.82
1-Octene	66.50	2.64	82.49	112.90	81.07	506.14	200.01	1.53
n-Decane	66.61	2.64	79.09	95.31	67.86	502.74	181.73	1.40
n-Octane	66.87	2.63	82.89	92.10	66.61	506.52	178.97	1.39
n-Tetradecane	66.20	2.63	84.24	88.65	64.24	507.89	175.18	1.36

TEA results of the 36 Solvents (Cont.)

Solvent	H_2/CO_2 --	H_2/CH_4 --	CAPEX-CO_2 Million $	OPEX-CO_2 Million $/year	LCOC $/ton.$CO_2$	CAPEX-H_2 Million $	OPEX-H_2 Million $/year	LCOH $/kg$H_2$
1-Hexanol	66.59	2.63	79.63	84.74	61.09	503.28	171.24	1.33
1-Nonanal	67.41	2.64	80.70	89.12	64.26	504.36	170.07	1.32
1-Octanol	58.25	2.64	81.69	89.64	64.81	505.35	175.65	1.36
DES	66.30	2.68	52.91	43.05	31.56	478.57	135.14	1.06
NBAc	63.70	2.64	50.26	37.84	27.84	476.36	130.35	1.04
NPAc	59.53	2.64	63.32	58.49	42.68	489.05	150.44	1.18
TBP	65.69	2.64	74.18	72.91	53.05	497.82	161.62	1.26
DMF	63.36	2.63	51.02	46.76	34.10	474.67	134.14	1.07
PN	60.58	2.64	70.91	74.47	53.71	494.55	160.96	1.25
cis-Decalin	65.73	2.64	75.94	78.38	56.68	499.59	165.16	1.29
Cyclohexanone	66.18	2.63	84.87	73.16	53.43	508.51	160.03	1.26
MNPh	66.07	2.63	77.32	84.28	60.98	500.97	170.27	1.32
NMP	62.61	2.63	68.40	67.53	49.09	492.05	155.04	1.22
PC	65.37	2.63	60.87	59.36	43.13	484.52	146.88	1.16
PEGDME	66.00	2.64	52.26	45.18	33.01	478.93	135.66	1.07
PEGPDMS-1	63.63	2.64	51.02	38.57	28.35	476.76	130.60	1.04
PEGPDMS-3	62.01	2.64	55.49	47.49	34.69	481.23	139.50	1.10

Development of Artificial Neural Network (ANN) in Matlab for the SMR-CCS

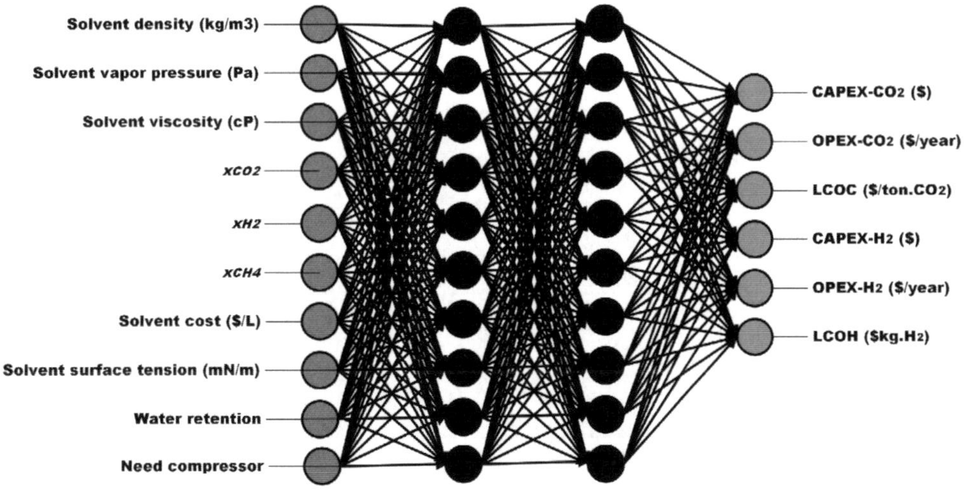

10 inputs

2 hidden layers

6 outputs

ANN Training

Training 36 physical solvents

Using Levenburg-Marquardt training method

Output	R^2 Training	AARE (%)
CAPEX-CO_2 capture	0.9999	0.21%
OPEX-CO_2 capture	0.9998	0.51%
LCOC	0.9967	1.16%
CAPEX-H_2 production	0.9999	0.03%
OPEX-H_2 production	0.9997	0.21%
LCOH	0.9979	0.38%

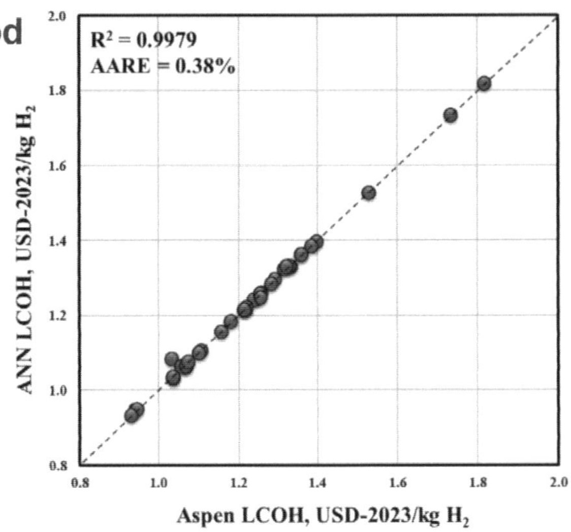

ANN Testing

Testing using Icosane and Tetraethylene Glycol as physical solvents

	Icosane ($C_{20}H_{42}$)		Tetraethylene Glycol ($C_8H_{18}O_5$)	
	Aspen	ANN	Aspen	ANN
CAPEX-CO_2 capture, MM 2023-$	83.72	78.77	70.72	67.42
OPEX-CO_2 capture, MM 2023-$/year	91.64	85.85	66.28	65.09
LCOC, 2023-$/ton CO_2	66.16	61.55	48.26	47.23
CAPEX-H_2, MM 2023-$	507.22	503.49	494.34	490.94
OPEX-H_2, MM 2023-$/year	198.86	190.17	173.30	171.29
LCOH, 2023-$/kg H_2	1.38	1.31	1.21	1.19
AARE	4.86%		1.99%	

Concluding remarks

- Aspen Plus V.12.1 was used to develop a modified process for CO_2 capture from the SMR-CCS for producing blue hydrogen.
- Aspen Custom Modeler V.12.1 was used to simulate H_2 membrane
- Aspen Adsorption V.12.1 was used to model for PSA
- 36 physical solvents were employed in a packed-bed absorber for more than 97 mol% CO_2 capture from the shifted fuel gas.
- CO_2 absorber hydraulics and mass transfer coefficients were calculated. k_L was controlling the overall mass transfer in the CO_2 absorber.

Concluding remarks (cont.)

- TEA was performed for the 36 solvents and showed that the BAC solvent (1AB-DECAM) had the lowest LCOH among all 36 solvents.
- The modified SMR-CCS was compared with DOE/NETL SMR-CCS process and found to be more advantageous due to its higher H_2/CO_2 and H_2/CH_4 ratios, and its lower CAPEX and LCOH.
- An ANN model was developed in Matlab, trained to predict the TEA results for the SMR-CCS using the 36 physical solvents. The ANN was able to predict accurately without using Aspen the TEA values for two other physical solvents which were not used in its development. Thus, the ANN is a powerful Machine Learning tool, providing huge timesaving for using physical solvents to produce blue H_2.

Disclaimer

Acknowledgement

This technical effort was performed in support of the US Department of Energy's ongoing research in Carbon Capture. The research was executed through the NETL Research and Innovation Center's Transformational Carbon Capture FWP under the RSS contract 89243318CFE000003.

References

1. Worldemissionclock. (2024). GHG emissions by sectors in 2024. Available: https://worldemissions.io/.

2. P. M. Kellstedt, S. Zahran, and A. Vedlitz, "Personal efficacy, the information environment, and attitudes toward global warming and climate change in the United States," Risk Analysis: An International Journal, vol. 28, no. 1, pp. 113-126, 2008.

3. E. S. Rubin, C. Chen, and A. B. Rao, "Cost and performance of fossil fuel power plants with CO2 capture and storage," Energy policy, vol. 35, no. 9, pp. 4444-4454, 2007.

4. Vitalie Ciobanu, "Hydrogen Economy." https://www.linkedin.com/pulse/hydrogen-economy-20-24022023-vitalie-ciobanu/

5. Haryanto, Agus, Sandun Fernando, Naveen Murali, and Sushil Adhikari. "Current status of hydrogen production techniques by steam reforming of ethanol: a review." Energy & Fuels 19, no. 5 (2005): 2098-2106.

6. E. Lewis et al., "Comparison of commercial, state-of-the-art, fossil-based hydrogen production technologies," National Energy Technology Laboratory (NETL), Pittsburgh, PA, Morgantown, WV, 2022.

7. Siefert, N.S., Narburgh, S. and Chen, Y., 2016. Comprehensive exergy analysis of three IGCC power plant configurations with CO2 capture. Energies, 9(9), p.669.

8. E. Perea and I. Jayaweera, "Development and testing of a high-temperature PBI hollow-fiber membrane technology for pre-combustion CO_2 capture," 2022.

9. Wales, M.D., Gebremichael, E., Wang, X., Perea, E., Jayaweera, P. and Jayaweera, I., 2021. Flue gas desulfurization (FGD) wastewater treatment using polybenzimidazole (PBI) hollow fiber (HF) membranes. Membranes, 11(6), p.430.

10. J.-G. Jee, M.-B. Kim, and C.-H. Lee, "Adsorption characteristics of hydrogen mixtures in a layered bed: binary, ternary, and five-component mixtures," Industrial & engineering chemistry research, vol. 40, no. 3, pp. 868-878, 2001.

11. J. Yang, C.-H. Lee, and J.-W. Chang, "Separation of hydrogen mixtures by a two-bed pressure swing adsorption process using zeolite 5A," Industrial & engineering chemistry research, vol. 36, no. 7, pp. 2789-2798, 1997.

12. Leva, M. (1992) 'Reconsider packed-tower pressure-drop correlations', Chemical engineering progress. American Institute of Chemical Engineers, 88(1), pp. 65–72.

13. R. Billet and M. Schultes, "Prediction of mass transfer columns with dumped and arranged packings: updated summary of the calculation method of Billet and Schultes," Chemical Engineering Research and Design, vol. 77, no. 6, pp. 498-504, 1999.

14. T. R. Brown, "Cost Engineering: Equipment Purchase Cost," *Chemical Engineering,* pp. 51-53, 2019.

15. Towler, Gavin, and Ray K. Sinnott. *Chemical engineering design: principles, practice and economics of plant and process design*. Elsevier, 2008.

16. https://toweringskills.com/financial-analysis/cost-indices/

U.S. DEPARTMENT OF ENERGY | NATIONAL ENERGY TECHNOLOGY LABORATORY | University of Pittsburgh

Performance of Anthracite Derived Activated Carbon in Adsorbing CO_2 across Different Gas Concentrations[1]

Zeyuan Tian [a], Gabriel Bonsall [a], Zheng Yao [a*], Tom Lowe [b], Greg Driscoll [b], Boyd Kreglow [b], Harold Schobert [c], and Carlos E. Romero [a*]

[a] *Energy Research Center, Lehigh University, 117 ATLSS Dr., Bethlehem, PA 18015, USA*

[b] *Blaschak Anthracite Corporation, 1166 W Centre St, Mahanoy City, PA 17948, USA*

[c] *Schobert International LLC, 22401 Hunter Ridge Circle, Jordan, MN 55352, USA*

Abstract

This study investigates the carbon dioxide (CO_2) adsorption performance of anthracite based activated carbon (ABAC) prepared via chemical activation using NaOH and K_2CO_3 across a range of concentrations. Anthracite obtained from Blaschak Anthracite Corp. underwent modification at temperatures spanning 500°C to 900°C with varying mass ratios of modification agents to carbon and was subsequently ground to particle sizes below 355 μm: these operations help to optimize surface area and porosity. CO_2 adsorption experiments were conducted at 25°C under three distinct conditions: direct air capture (0.04% CO_2), point source capture with lower

[1] This paper was presented at 41st annual international Pittsburgh energy and carbon management conference

concentration (4% CO_2), and point source capture with higher concentration (13% CO_2) environments. The study identified the optimal ABAC formulation achieving a CO_2 adsorption capacity of 3.63 mmol/g at 1 bar and 0 °C following treatment with NaOH. Comparison with commercial granular activated carbon (GAC) BPL showed that ABAC exhibited either comparable or superior performance, with adsorption percentages of 25.81% and 16.80% in lower and higher concentration environments, respectively, compared to BPL's 18.79% and 9.95%. Evaluation metrics such as sustained adsorption time, effective adsorption period, total adsorbed concentration, and breakthrough concentrations further supported ABAC's efficacy, highlighting its suitability for applications requiring efficient CO_2 capture, especially under higher concentration conditions. These findings underscore ABAC's potential as an effective and versatile sorbent for mitigating CO_2 emissions in diverse environmental and industrial settings.

1. Introduction

Carbon dioxide is Earth's most important greenhouse gas: a gas that absorbs and radiates heat. Unlike oxygen or nitrogen (which make up most of our atmosphere), greenhouse gases absorb heat radiating from the Earth's surface and re-release it in all directions—including back toward Earth's surface [1]. CO_2 capture technologies for large point sources are vital in mitigating the environmental impact of industrial emissions of CO_2. Various methods exist for CO_2 removal, including pre-combustion decarbonization [2], oxy-fuel combustion [3], chemical looping combustion [4], and post-combustion capture [5]. However, the development of a cost-effective process with low energy requirements and high CO_2 capture efficiency remains an ongoing challenge. Researchers continue to explore solutions that balance these factors to create scalable and economically viable technologies for large-scale CO_2 sequestration.

The injection of dry sorbents upstream of particulate control devices offers substantial potential for controlling gas-phase CO_2 emissions from utility boilers. However, dry sorbent injection treatments for CO_2 capture vary significantly depending on the characteristics of the sorbent, injection method, CO_2 concentrations, flue gas composition, and operating conditions. Municipal solid waste (MSW) incinerators and coal-fired plants are recognized as important CO_2 emission sources. The flue gas from coal combustion contains CO_2, while other byproducts, such as sulfur dioxide (SO_2) and nitrogen oxides (NO_x), can also influence CO_2 capture efficiency.

Several parameters, including CO_2 is its own species in the flue gas, its composition, and process conditions like temperature, could impact sorbent performance during CO_2 capture. Studies have demonstrated that both physical and chemical adsorption methods can be used for CO_2 capture, with chemical adsorbents such as amine-based sorbents being more efficient in capturing CO_2 from flue gas streams than physical sorbents [6] . Different types of sorbents exhibit varying adsorption behaviors depending on CO_2 concentration and flue gas conditions [7] . Moreover, gas species such as SO_2 and NO_x can influence CO_2 adsorption performance. Furthermore, process factors such as temperature variations and sorbent residence time within the power plant must be considered, as these parameters can significantly affect adsorption efficiency [8].

In addition to these factors, the physical properties of sorbent materials are critical in determining their performance for CO_2 capture. Activated carbon (AC) is one of the most promising dry sorbents for CO_2 removal due to its excellent surface properties and cost-effectiveness [9]. Chemical activation processes, particularly those using amines or sodium/potassium hydroxide, are widely employed to enhance the performance of AC for CO_2 adsorption. However, the high cost of chemical preparation is a significant limitation. Therefore, the development of novel, affordable, and highly effective CO_2 sorbents is of great interest. Anthracite-based activated carbon (ABAC) has shown potential as a low-cost alternative for CO_2 adsorption due to its microporous structure and high surface area. Anthracite, a rank of coal, is particularly promising for this purpose because of its high carbon content, low impurities, and favorable pore structure [10]. However, the mechanisms governing CO_2 adsorption on AC and ABAC are not yet fully understood. Variables such as pore size distribution, surface functional groups, and the presence of impurities can influence the adsorption behavior.

To better understand the performance of ABAC for CO_2 capture, Lehigh University developed a laboratory-scale packed-bed experiment to evaluate sorbents at different carbon dioxide concentration environments. Preliminary results indicate that ABAC shows excellent performance for CO_2 capture in simulated flue gas conditions. Additionally, modifying ABAC with chemical impregnation, such as sodium hydroxide, significantly enhances its CO_2 adsorption capacity compared to traditional AC. This process offers a cost-effective alternative to other commercially available sorbents. The use of anthracite as raw material also opens up possibilities for developing high-performance sorbents with reduced abrasion losses and lower production costs.

In this study, ABAC was selected for its superior performance in controlling CO_2 emissions, simulating the natural gas combined cycle (NGCC), coal-fired power plant, and ambient air, without incurring additional processing costs. A comprehensive experimental design was conducted, including a slipstream setup and laboratory-scale tests to validate the effectiveness of ABAC for CO_2 capture from air or power sectors. Additionally, several commercially available activated carbons (CACs) were tested for comparison.

2. Methodology

This section outlines the comprehensive approach taken to prepare anthracite-based activated carbon (ABAC) for CO_2 capture and describes the procedures for evaluating its adsorption performance across different environmental conditions. The methodology is divided into three main parts: materials and procedures, experimental design, and the thermal regeneration process. Each part provides a detailed account of how the material was processed, tested, and analyzed to assess its effectiveness as a CO_2 adsorbent.

2.1 Materials and Procedures

The preparation of ABAC was designed to optimize the carbon porosity structure for high CO_2 adsorption capacity. The process involved selecting appropriate materials, activating the anthracite to enhance porosity, and refining the material to ensure uniformity and efficiency in adsorption tests.

2.1.1 Materials and Equipment

The main precursor for the activated carbon used in this study was anthracite supplied by Blaschak Anthracite, headquartered in Mahanoy City, Pennsylvania, a leading anthracite producer known for its innovative mining practices and environmental stewardship; its strong partnership with Lehigh University's Energy Research Center underscores its commitment to advancing sustainable energy solutions through cutting-edge research and technology.

Specifications	COARSE BLEND	BUCK	RICE	BARLY NO. 4 BLEND	NO. 5 BUCKWHEAT
Screen Size					
Inches	2 - 5/8	5/8 - 3/8	3/8 - 3/16	3/16 - 3/64	3/64 - 100mesh
mm	50-16	16-10	10-5	5-1	<1
Proximate Analysis (Dry Basis)					
% Ash	9.50	8.00	8.00	10.50	13.50
%Volatile	4.00	4.00	4.50	5.00	5.50
% Fixed Carbon	86.50	88.00	87.50	84.50	81.00
Ultimate Analysis (Dry Basis)					
% Carbon	85.65	87.40	86.85	83.75	80.15
%Hydrogen	1.80	1.70	1.75	1.80	2.00
% Nitrogen	1.20	1.30	1.30	1.50	1.60
% Sulfur	0.65	0.65	0.65	0.65	0.65
% Ash	9.5	8.0	8.0	10.5	13.5
% Oxygen	1.20	0.95	1.45	1.80	2.10
Heating Value (Dry Basis)					
Btu/lb	13,300	13,400	13,400	13,000	12,500
Kcal/kg	7,390	7,450	7,450	7,230	6,950
Ash Fusion Temp. (Softening Point)					
°F	2,700+	2,700+	2,700+	2,700+	2,700+
°C	1,480+	1,480+	1,480+	1,480+	1,480+
Moisture (As Received)					
%	5.00	7.00	7.00	10.00	15.00
Hardgrove Grindability Index	38	38	38	40	40

Figure 1 Blaschak Anthracite Properties [11]

Anthracite is particularly suitable for generating activated carbon due to its dense structure and the minimal presence of contaminants, as seen in figure 1, which allows for more controlled chemical activation. By using high-quality anthracite, the final ABAC product can achieve a high level of purity, stability, and structural integrity [10], all of which are essential for effective CO_2 capture.

Chemical activation of the anthracite was conducted using sodium hydroxide (NaOH) and potassium carbonate (K_2CO_3). These agents were selected for their ability to create a porous structure within the carbon material [12]. NaOH and K_2CO_3 are commonly used in the activation process because they react with the organic content in coal and forming intricate pore structures. This process increases the surface area and creates micro- and mesopores that facilitate gas adsorption [13]. The choice of these chemicals was based on their proven effectiveness in generating high porosity activated carbon suitable for capturing gaseous pollutants, including CO_2 [12].

Using these two chemical agents, various samples of ABAC were prepared for adsorption testing. These samples were evaluated alongside the commercial activated carbon BPL (4x6), supplied by Calgon Corp, which served as a benchmark, the specifications are illustrated in Figure 2. This comparison aimed to assess the performance of the ABAC samples relative to an established industry standard.

Specifications	BPL 4x6
Butane Activity, wt%	23.3 (min)
Moisture (As Packaged), wt%	2 (max)
Hardness Number	95 (min)
Density (Apparent), g/cc	0.43 (min)
Mean Particle Diameter, mm	3.7 (min)
4 US Mesh [4.75mm], wt%	15 (max)
< 7 US Mesh [2.80mm], wt%	8.0 (max)

Figure 2 BPL commercial granular activated carbon Specifications [14]

Table 1 presents the experimental conditions applied to two groups of samples treated with different chemical agents, alongside the commercial activated carbon used for comparison. The table outlines key parameters such as mass ratio, activation temperature, duration, providing a comprehensive overview of the preparation conditions for each sample group.

Table 1 Experimental conditions applied to different samples

Sample name	$NaOH$ #2	K_2CO_3 #3	BPL
Ratio	4:1	1:1	/
Furnace time	1hr	5hr	/
Furnace Temp	800°C	700°C	/
size	<355um	<355um	<355um

The laboratory setup utilized for the study comprised a tube furnace for chemical activation, a ball mill for particle size reduction, and high-precision CO_2 sensors for performance evaluation. The tube furnace, depicted schematically in Figure 3, was central in creating the controlled high-temperature environment essential for the activation process. Its integrated programming system enabled precise regulation of parameters such as activation duration and temperature, as outlined in Table 1. Furthermore, the system allowed for adjustable flow rates, tailored to accommodate various activating agents. Depending on the selected agent, specific carrier gases were introduced to ensure optimal reaction conditions.

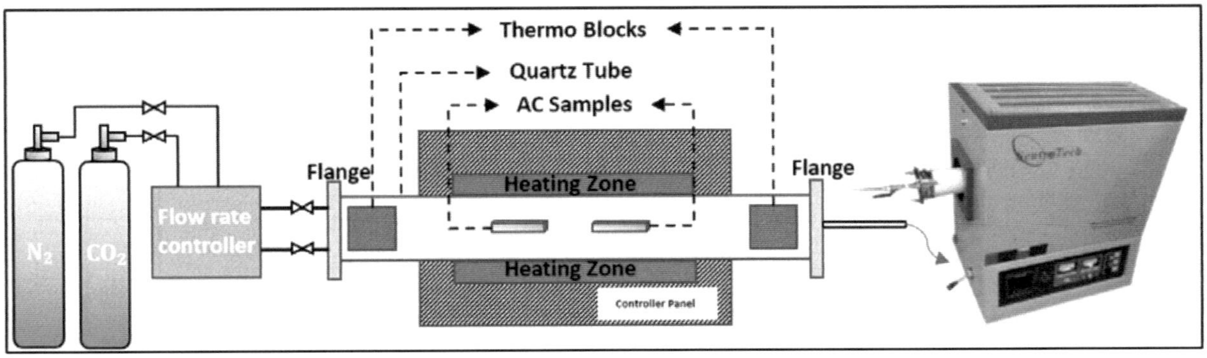

Figure 3 Scheme of Tube Furnace in Thermal Lab at Lehigh University's Energy Research Center

The ball mill was employed to finely grind the material, thereby maximizing its surface area for enhanced CO_2 capture. As illustrated in Figure 4, the grinding process utilized an 80 mL stainless steel container, which rotated with the samples and grinding media to achieve uniform particle size reduction. To ensure consistency and eliminate size-related variables, identical grinding balls—10 pieces of 8 mm zirconia—were used with all samples. Additionally, the milling parameters, including motor speed and grinding duration, were maintained constant for all experiments, ensuring comparability of results.

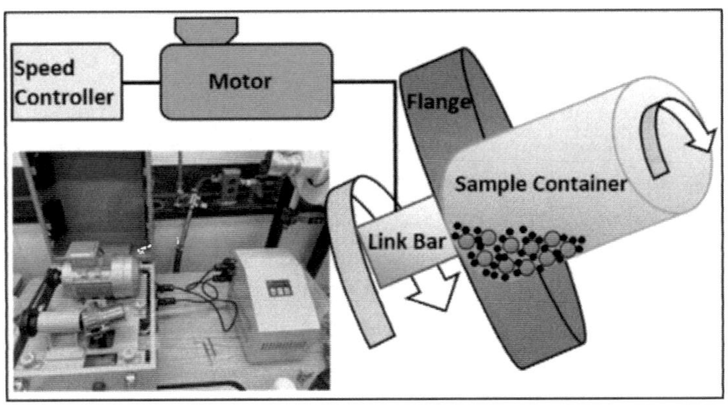

Figure 4 Scheme of Ball Mill in Machine Lab at Lehigh Energy Research Center

High-precision CO_2 sensors were utilized to monitor adsorption performance, chosen specifically for their sensitivity and accuracy across varying concentration ranges. The study employed two models sourced from CO2Meter: the T6713 Miniature CO_2 sensor and the K33 BLG 30% CO_2 sensor, each tailored for distinct operational stages. Figure 5 provides a detailed overview of these sensors, highlighting their respective key features and specifications.

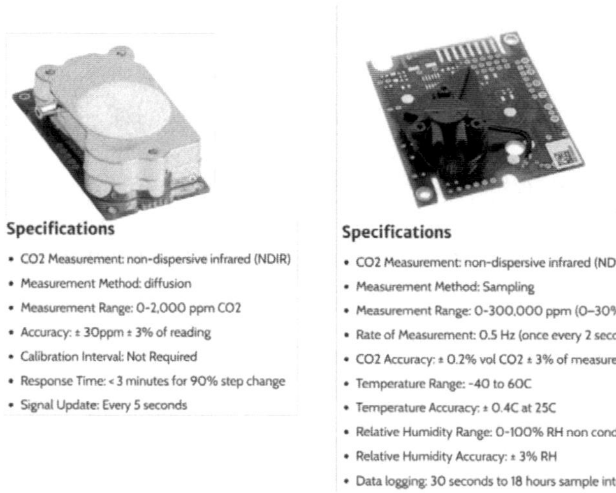

Figure 5 T6713 Miniature sensor (left) & K33 BLG 30% sensor(right) from CO2METER [15]

2.1.2 Procedures

The preparation and testing of anthracite based activated carbon LU-ABAC followed a rigorous and systematic approach, aimed at maximizing its CO_2 adsorption performance and benchmarking it against commercial activated carbon (LU-AC).

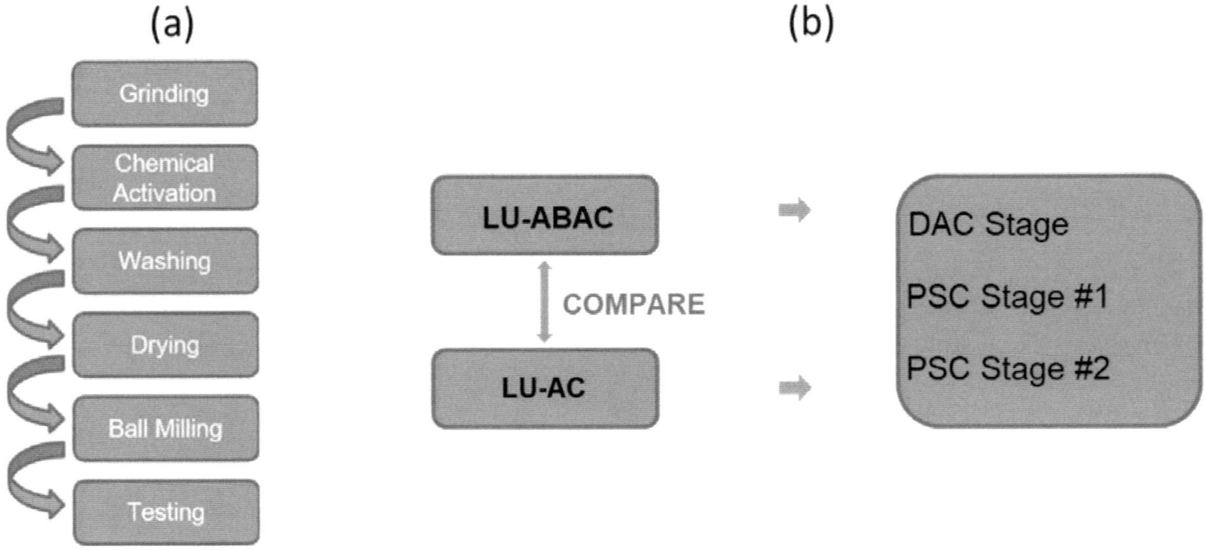

Figure 6(a) Scheme of the procedure for activating carbon; 6(b) Scheme of the lab-scale adsorbing tests

As the Figure 6 (a) shows, the preparation process began with grinding, where large coal particles were broken down into smaller sizes to increase surface area and ensure uniformity in subsequent steps. This was achieved using industrial grinding equipment, ensuring the material was fine enough to facilitate chemical reactions during activation.

These ground samples were then chemically activated by mixing with either NaOH or K_2CO_3, followed by high-temperature treatment in a tube furnace under controlled atmospheres of N_2 or CO_2. The activation process was precisely regulated, with parameters such as temperature, duration, and flow rates adjusted to optimize porosity and adsorption capacity.

Once activated, the carbon samples underwent a washing stage using hydrochloric acid (HCl) to remove residual activating agents and impurities. The cleaned samples were then dried to eliminate moisture and ensure stability. Ball milling was subsequently performed to reduce the carbon into fine powders, significantly increasing their surface-to-volume ratio for improved adsorption efficiency.

Once prepared, the activated carbon samples underwent a series of laboratory-scale CO_2 adsorption tests to evaluate their performance. The tests were conducted in distinct stages depict in Figure 6 (b), including DAC (Direct Air Capture) and PSC (Point Source Capture) scenarios, specifically PSC with 4% CO2 for NGCC, and PSC with 13% CO2 for coal fired power plants. These stages simulated different CO_2 concentration environments to assess the versatility and efficiency of the materials under varying conditions. The LU-ABAC samples were compared directly with the commercial LU-AC to determine their adsorption capabilities.

The results demonstrated that the LU-ABAC samples exhibited comparable performance to the commercial LU-AC, highlighting their potential as a viable material for CO_2 capture applications. The systematic and detailed preparation process, coupled with rigorous testing, underscores the importance of each step in developing high-performance activated carbon. By optimizing every stage—grinding, chemical activation, washing, drying, ball milling, and testing—this study provides a robust framework for producing and evaluating activated carbon materials for industrial and environmental applications.

2.2 Experimental Design

The experimental design was developed to test ABAC's adsorption capabilities under different CO_2 concentration conditions, providing insights into its performance across scenarios from ambient air to industrial emissions. The tests were structured in three stages, each representing a distinct CO_2 concentration and intended to assess the material's adaptability to various capture applications.

2.2.1 Directed Air Capture (DAC) stage

The first stage of testing focused on assessing ABAC's performance in capturing CO_2 from ambient air. This low-concentration setting simulates the conditions of direct air capture (DAC), where CO_2 levels are approximately 400 ppm. The aim was to determine whether ABAC could effectively adsorb CO_2 in conditions that closely resemble the natural atmosphere, as DAC applications often target the removal of CO_2 from the open air.

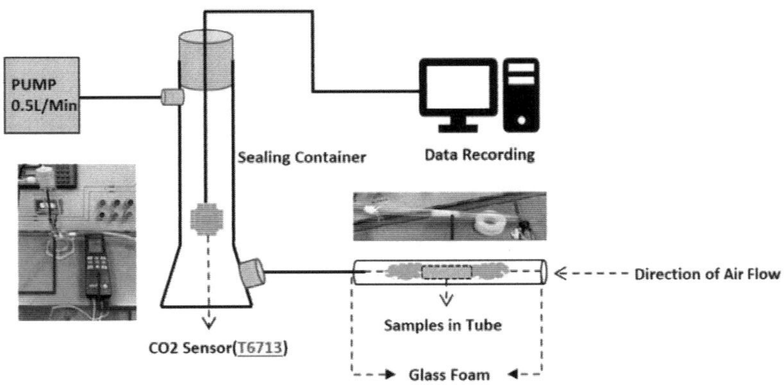

Figure 7 Scheme of the Directed Air Capture (DAC) Stage

The schematic in figure 7 depicts a controlled air flow system for testing CO_2 adsorption. A pump delivers air at 0.5 L/min through a sealed container housing the sample material secured by glass foam. The air passes through the sample tube, and the ABAC samples were exposed to ambient air in a controlled chamber, where the CO_2 concentration was continuously monitored using a T6713 CO_2 sensor downstream. Data is recorded and analyzed via a connected computer, ensuring accurate and repeatable measurements. The decrease in CO_2 levels over time was recorded, enabling the calculation of adsorption capacity at atmospheric concentrations. The DAC stage is important for applications involving environmental air filtration, where CO_2 capture from low concentrations is essential for climate-related interventions.

2.2.2 Point Source Capture (PSC) Stage

The point-source capture (PSC) testing involved two CO_2 concentrations shown in Figure 8, designed to evaluate the performance of anthracite-based activated carbon (ABAC) under varying industrial CO_2 emission conditions using a consistent experimental setup. The system included separate gas cylinders containing 4% and 13% CO_2 concentrations, each connected to a flow rate controller to precisely regulate the gas flow. The gas was directed through valves into a glass tube packed with ABAC samples, which were sealed with glass foam to maintain a controlled environment. At the tube's outlet, a K33 BLG CO_2 sensor was installed to monitor CO_2 levels, with data continuously recorded by a computer for detailed analysis.

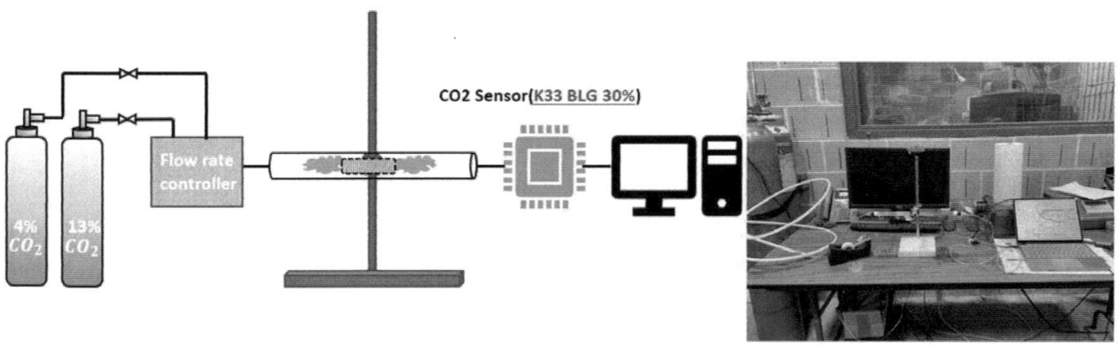

Figure 8 Scheme of the Point Source Capture (PSC) Stage

The 4% CO_2 concentration, representative of emissions from natural gas power plants, was used to simulate moderate industrial emissions. This stage assessed ABAC's capacity for CO_2 capture under conditions aligning with regulatory standards for natural gas-based power generation. The K33 BLG sensor, known for its accuracy of ±3% at moderate CO_2 levels, recorded data at 30-second intervals, enabling an in-depth examination of adsorption trends over time.

The higher 13% CO_2 concentration presented a more challenging environment – typical of emissions from coal-fired power plants. This stage tested ABAC's ability to perform under high-emission conditions where rapid and efficient CO_2 capture is crucial. The same K33 BLG sensor, calibrated for high CO_2 concentrations, ensured precise measurements. The data from this stage provided valuable insights into ABAC's performance under demanding conditions, demonstrating its potential for use in high-emission industries that require robust, high-capacity adsorbents. Collectively, these two CO_2 concentration tests allowed for a comprehensive assessment of ABAC's effectiveness in a range of industrial emission scenarios.

2.3 Thermal Regeneration Design

To understand the reusability and longevity of ABAC, a thermal regeneration process was incorporated following each adsorption cycle. Thermal regeneration is a crucial step that involves heating the activated carbon to release the adsorbed CO_2, restoring its adsorption capacity for subsequent cycles. This process mimics the real-world demands placed on adsorbents, where materials need to maintain their efficiency over repeated use in industrial and environmental settings.

In this study, thermal regeneration was carried out in a controlled environment, one group of the comparable samples were made in the Table 2, where the ABAC was heated to a specific temperature that facilitated the release of CO_2 without compromising the carbon's structural integrity. By releasing the adsorbed CO_2, the regeneration process liberated adsorption sites within the carbon matrix, allowing the material to be reused for further CO_2 capture. This cyclic process was repeated over multiple adsorption and desorption cycles to simulate long-term use.

Table 2 Regeneration conditions applied to NaOH and BPL

Sample	NaOH #2	BPL
Gas flow	N_2	N_2
Furnace time	0.5 hr	0.5 hr
Furnace Temp	500°C	500°C
Used for	13% PSC	13% PSC

The regeneration tests were carefully monitored to assess any changes in the adsorption capacity or structural stability of the ABAC. Each regeneration phase was followed by additional adsorption tests to measure the material's capacity and adsorption rates after repeated use. These tests are essential for determining whether ABAC can sustain its performance over time, which is critical for applications in continuous or cyclic operations. The outcomes from the thermal regeneration phase are key to evaluating the feasibility of deploying ABAC as a long-term CO_2 adsorbent in practical applications.

2.4 Measurements and Parameters Definition

After recording the data with the CO_2 sensor during the adsorption process, a graph in Figure 9 depicting CO_2 concentration over time is generated to analyze the performance of ABAC in CO_2 capture. This graph provides critical parameters that define the adsorption efficiency and behavior of the material.

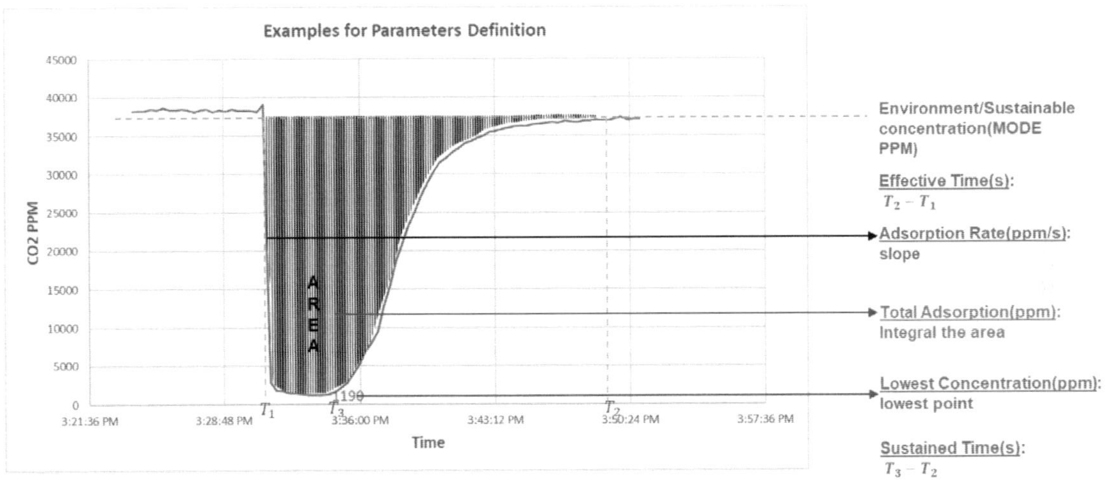

Figure 9 Results from the data recording and the parameters definition

The first key parameter is the percent reduction, which measures the percentage of CO_2 removed during the adsorption process. It is calculated by comparing the initial CO_2 concentration at the start of adsorption (T_1) with the lowest concentration (C_{lowest}) reached during the process. This reduction illustrates the effectiveness of the ABAC in capturing CO_2 from the gas flow.

The graph also identifies the effective time, which is the duration between T_1(the beginning of adsorption) and T_2(when the system reaches equilibrium). This time interval reflects how quickly the ABAC achieves maximum adsorption under the given conditions.

Another critical parameter is the adsorption rate, representing the speed at which CO_2 is adsorbed during the effective phase. This rate is determined by calculating the slope of the CO_2 concentration curve, which is derived from the change in concentration (ΔC) over time (Δt).

The total adsorption is evaluated by integrating the area under the concentration curve between T_1 and T_2.This parameter quantifies the total amount of CO_2 captured by the ABAC sample, offering a direct measure of its adsorption capacity.

Additionally, the graph indicates the lowest concentration, which corresponds to the minimum CO_2 level recorded during the test. This point highlights the maximum adsorption performance achieved by the material. Following the equilibrium phase, the sustained time, measured from T_2 to T_3, reflects how long the system maintains stable adsorption conditions after equilibrium is reached. This detailed analysis of the graph provides a comprehensive understanding of ABAC's performance under varying CO_2 concentrations, emphasizing its potential for industrial carbon capture applications.

Also, from this graph, several key equations are applied to extract meaningful insights into the adsorption performance of anthracite-based activated carbon (ABAC). Each equation plays a critical role in evaluating how efficiently and effectively CO_2 is captured, providing a clear picture of the material's potential in industrial applications.

The **Percent Reduction** equation measures the proportion of CO_2 removed during the effective adsorption phase. It is calculated as:

$$Percent\ Reduction = \left(\frac{C_{initial} - C_{lowest}}{C_{initial}}\right) \times 100$$

Here, $C_{initial}$ is the initial CO_2 concentration at the start of adsorption (T_1), and C_{lowest} is the lowest concentration observed during the process. This calculation is important because it quantifies how much CO_2 is effectively captured, serving as a direct indicator of ABAC's efficiency in reducing emissions.

The **effective time**, represented by the difference $T_2 - T_1$, determines the duration required for the adsorption process to reach equilibrium. This parameter helps identify how quickly the material can achieve stable adsorption, which is critical for time-sensitive industrial processes.

To evaluate the speed of the adsorption process, the **adsorption rate** equation is used:

$$Adsorption\ Rate = \frac{\Delta C}{\Delta t}$$

Here, ΔC is the change in CO_2 concentration, and Δt is the corresponding change in time. This rate represents how rapidly ABAC reduces CO_2 concentration during the effective phase. A higher rate indicates better kinetics, which is crucial for applications requiring rapid response to high emission rates.

The **total adsorption** equation measures the cumulative amount of CO_2 adsorbed over the effective phase, expressed as the area under the concentration curve between T_1 and T_2:

$$Total\ Adsorption = \int_{T_1}^{T_2} C(t)dt$$

These integral captures the overall adsorption capacity of ABAC, highlighting its ability to handle large quantities of CO_2. This parameter is critical for assessing whether the material can meet industrial demands for emission control.

The **lowest concentration** identifies the minimum CO_2 concentration C_{lowest} reached during the adsorption process. This value represents the most effective point of adsorption and reflects ABAC's maximum performance under test conditions.

Finally, the **sustained time**, calculated as $T_3 - T_2$, assesses how long the CO_2 concentration remains stable after equilibrium is achieved. This parameter is important for evaluating the long-term stability of ABAC in maintaining low CO_2 levels, which is vital for continuous industrial operations.

3.Results and Discussion

3.1 Direct Air Capture

The DAC tests for ABAC treated with different chemicals show varied effectiveness in CO_2 capture. The graphical results shown in Figure 10 demonstrate how each treatment modifies the ABAC's performance. The BPL sample starts at 428 PPM and drops to 299 PPM, showcasing a significant reduction in CO_2 levels. The NaOH #2 treated ABAC, beginning at 536 PPM, decreases to 324 PPM, indicating its capacity to handle higher initial CO_2 concentrations effectively. Conversely, the K_2CO_3 #3 treatment shows a smaller decrease from 438 PPM to 419 PPM, highlighting its lower effectiveness in this test.

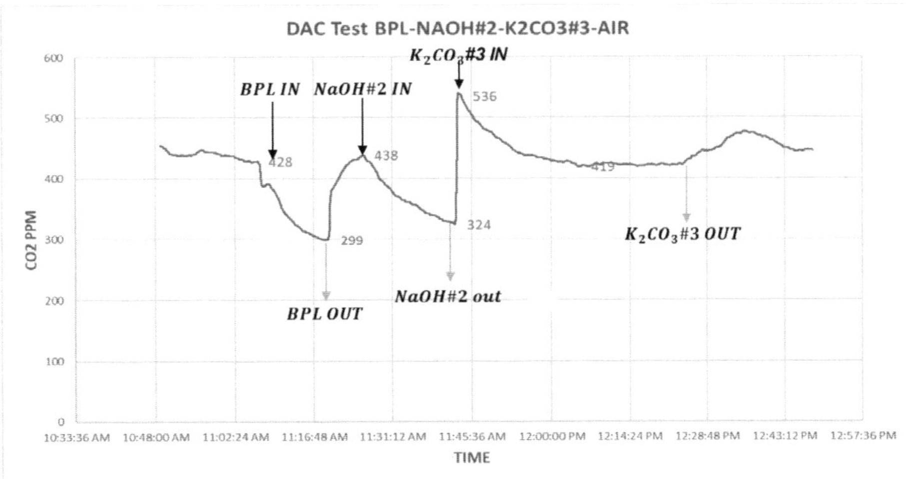

Figure 10 DAC Test Graph

From results in table 3, BPL shows a notable 17.55% reduction in CO_2, the highest among the samples. Although NaOH #2 has a lower percent reduction at 10.71%, its total adsorption is substantial, suggesting it is effective over longer periods. This feature is particularly useful for continuous CO_2 capture applications. K_2CO_3 #3 has the least effectiveness with an 8.90% reduction, showing its limited capacity under the test conditions.

Table 3 Important Parameters Table for DAC Test Results

Sample	NaOH #2	K_2CO_3 #3	BPL	*Glass Wool*
Mass (g)	*0.5*	*0.5*	*0.5*	/
Particle Size	*<355um*	*<355um*	*<355um*	/
Adsorption Rate (ppm/s)	0.11	0.05	0.18	~0
Sustained Time (s)	992	980	727	20
Effective Time (s)	3100	1653	6420	50
Total Adsorption (ppm)	$>1.41 \times 10^5$	$\sim 6.21 \times 10^4$	$> 4.75 \times 10^5$	~1.33
Percent Reduction	10.71%	8.90%	17.55%	~0
Lowest Concentration (ppm)	324	419	299	445

An important part of the experiment includes a control test using glass wool, which essentially shows no reduction in CO_2, confirming that the adsorption observed in other samples is due to the ABAC and not influenced by the glass wool or the sealed system. This control setup helps verify that the CO_2 reductions are indeed attributable to the treated ABAC's properties and not external factors.

Overall, these results clarify how each chemical treatment affects the ABAC's ability to capture CO_2. While BPL is efficient for quick reduction, NaOH #2 is advantageous for longer, sustained CO_2 management. The less-effective K_2CO_3 #3 might require further optimization for better performance. The inclusion of the control test ensures the reliability of the observed effects, excluding any influence from the experimental setup itself.

3.2 Point Source Capture in Natural Gas Power Plant Environment

In the 4% CO_2 PSC test, which simulate the natural gas power plant environment, different treatments of ABAC were evaluated for their effectiveness in capturing CO_2. The results show each sample's dramatic reduction from initial CO_2 levels to their respective lowest concentrations. For example, NaOH #2 starts high and drops significantly to 840 PPM, indicating a substantial decrease.

Table 4 Important Parameters Table for PSC#1 Test Results

Sample	NaOH #2	K_2CO_3 #3	BPL	Glass Wool
Mass (g)	2	2	2	2
Particle Size	<355um	<355um	<355um	/
Adsorption Rate (ppm/s)	457.13	1812.00	603.67	490.50
Sustained Time (s)	80	40	80	40
Effective Time (s)	618	481	946	40
Total Adsorption (ppm)	6.52×10^6	2.28×10^6	1.34×10^7	1.70×10^5
Percent Reduction	28.15%	12.67%	37.87%	11.35%
Lowest Concentration (ppm)	840	1170	1190	27600

The performance data in table 4 provides a clear view of each treatment's efficiency:

- NaOH #2 is notable with an adsorption rate of 457.13 ppm/s and a total adsorption of 6.52×10^6 ppm, effectively capturing CO_2 over 618 seconds and reducing the concentration by 28.15%.
- K_2CO_3 #3 records the fastest adsorption rate at 1,812.00 ppm/s but achieves a smaller percent reduction of 12.67%, suggesting it processes CO_2 quickly but not as thoroughly.
- BPL shows the most significant reduction at 37.87%, effectively lowering the CO_2 to 1,190 PPM over 946 seconds, demonstrating its capacity to sustain CO_2 capture efficiently.

Still, the control test using glass wool, showing an 11.35% reduction to a high of 27,600 PPM, is crucial as it ensures the reductions seen in the ABAC samples are solely due to the carbon's properties, not influenced by the glass wool or the experimental setup's sealing.

These findings underscore the distinct capabilities of each ABAC treatment under a 4% CO_2 concentration, highlighting NaOH #2's superior performance in speed and overall CO_2 absorption, with BPL also showing robust, long-term reduction capabilities. This information is vital for optimizing ABAC treatments for applications requiring effective and sustained CO_2 management.

3.3 Point Source Capture in Coal Fired Power Plant Environment

In the 13% CO_2 PSC test, similarity tests were made to simulate the adoption ability in the coal fired power plant environment. The test results, displayed in table 5, provide insights into each treatment's effectiveness at this higher concentration.

Table 5 Important Parameters Table for PSC#2 Test Results

Sample	NaOH #2	K_2CO_3 #3	BPL	*Glass Wool*
Mass (g)	2	2	2	2
Particle Size	<355um	<355um	<355um	/
Adsorption Rate (ppm/s)	2115.33	1647.00	2101.67	1215.50
Sustained Time (s)	100	40	100	40
Effective Time (s)	609	474	209	86
Total Adsorption (ppm)	8.85×10^6	1.20×10^6	4.85×10^6	6.06×10^5
Percent Reduction	11.34%	1.97%	18.05%	5.49%
Lowest Concentration (ppm)	1020	95000	1840	103630

The result highlights significant drops in CO_2 levels for each treatment, showing the initial high concentrations and their respective lowest points reached after treatment. For example, the NaOH #2 treatment starts at a peak and drops sharply to 1,020 PPM. This dramatic reduction showcases its effectiveness in capturing CO_2 from high concentration environments.

The table provides a clear comparison of key performance metrics across different samples:

- NaOH #2 achieves a high adsorption rate of 2,115.33 ppm/s and a total adsorption of 8.85 x 10^6 ppm. It sustains this performance over 609 seconds, reducing the CO_2 concentration by 11.34% to 1,020 PPM.
- K_2CO_3 #3, although having a substantial adsorption rate of 1,647.00 ppm/s, shows a lower percent reduction of 1.97%, and its lowest concentration reached is significantly higher at 95,000 PPM.
- BPL demonstrates a robust percent reduction of 18.05%, effectively lowering CO_2 to 1,840 PPM over an effective time of 209 seconds.

The control test using glass wool indicates a 5.49% reduction, resulting in a high ending concentration of 103,630 PPM. This inclusion confirms that the observed CO_2 reductions are genuinely attributable to the ABAC treatments and not influenced by the experimental setup, such as the glass wool or the sealed system's efficiency.

These results underline the unique capabilities of each ABAC treatment under a 13% CO_2 concentration scenario. NaOH #2 shows a superior performance in terms of speed and overall CO_2 absorption, making it particularly effective for high concentration environments. BPL also stands out for its high reduction percentage, proving its efficacy in sustained CO_2 management. This comprehensive analysis aids in understanding how each ABAC treatment can be optimized for specific applications where high CO_2 capture is crucial.

3.4 Thermal Regeneration for 13% PSC

In the thermal regeneration test conducted at 13% CO_2 concentration, we evaluated the efficacy of reactivated ABAC samples treated with NaOH, alongside the original BPL and NaOH #2 samples. This test aimed to determine the performance of ABAC after undergoing a thermal regeneration process, measuring how well the samples could recover their CO_2 adsorption capabilities.

The results, in table 6, show the CO_2 concentration over time for the reactivated samples, indicates significant reductions in CO_2 levels as the air passes through each sample, showcasing their recovery and effectiveness post-regeneration. For example, the reactivated BPL sample dramatically drops from an initial high to a lowest concentration, effectively demonstrating its renewed adsorption capacity.

Table 6 Important Parameters Table for Thermal Regeneration Test Results

Sample	Reactivated $NaOH$ #2	$NaOH$ #2	Reactivated BPL	BPL
Mass (g)	*2*	*2*	*2*	*2*
Particle Size	*<355um*	*<355um*	*<355um*	*<355um*
Adsorption Rate (ppm/s)	2104.8	2115.33	2007.2	2101.67
Sustained Time (s)	50	100	100	100
Effective Time (s)	450	609	1100	209
Total Adsorption (ppm)	8.41×10^6	8.85×10^6	4.40×10^7	4.85×10^6
Percent Reduction	35.96%	11.34%	25.39%	18.05%
Lowest Concentration (ppm)	680	1020	1350	1840

The table breaks down the performance of each sample in detail:

- Reactivated NaOH #2 and Reactivated BPL both show a robust return in functionality, achieving high adsorption rates and substantial total adsorption figures post-regeneration. For instance, reactivated NaOH #2 has an adsorption rate of 2,115.33 ppm/s, with a total adsorption reaching up to 8.85 x 10^6 ppm, showing it effectively reduces the CO_2 concentration to as low as 1,020 PPM.
- Comparatively, NaOH #2 and BPL in their original states before regeneration also maintain good performance metrics but with notable differences in their recovery capacity demonstrated by the reactivated samples.

The control in this test, highlighted by the glass wool sample, shows minimal adsorption capacity, ensuring that the observed reductions in the ABAC samples are attributable to the carbon material itself and not influenced by other factors in the setup.

These results are crucial as they illustrate the durability and reusability of ABAC after thermal regeneration, highlighting the potential for long-term use in industrial applications requiring frequent regeneration to maintain effective CO_2 capture. The data confirms that both NaOH treated, and baseline ABAC can regain a significant portion of their adsorption capacity, which is vital for their sustainable use in practical applications.

3.5 Accelerated Surface Area and Porosity (ASAP) and Scanning Electron Microscopy (SEM) Results

The results from the Accelerated Surface Area and Porosity (ASAP) and Scanning Electron Microscopy (SEM) analyses provide critical insights into the structure and functionality of Lehigh University's anthracite-based activated carbon (LU-ABAC) treated with NaOH.

The ASAP analysis quantitatively measures the surface area and porosity of the activated carbon, which are critical factors in determining its CO_2 adsorption capacity. For LU-ABAC treated with NaOH, the CO_2 adsorption capacity is reported at 3.63 mmol/g. This value is significantly higher compared to the commercial activated carbon (BPL) which has an adsorption capacity of 1.71 mmol/g. The enhanced capacity of the NaOH-treated ABAC indicates that the chemical treatment effectively increases the porosity and surface area, thereby improving the material's ability to capture more CO_2.

The SEM images, in figure 11, further elucidate the microstructural changes induced by the NaOH treatment. The images reveal a detailed view of the internal structure of LU-ABAC, highlighting the presence of macro-, meso-, and micropores. These images visually demonstrate that NaOH treatment leads to the development of a more complex pore structure, with a significant increase in both the number and size of pores across different scales. The labeled SEM image also identifies regions of activated carbon matrix along with the molecules of adsorbate, which are likely CO_2 molecules, showing where the adsorption takes place.

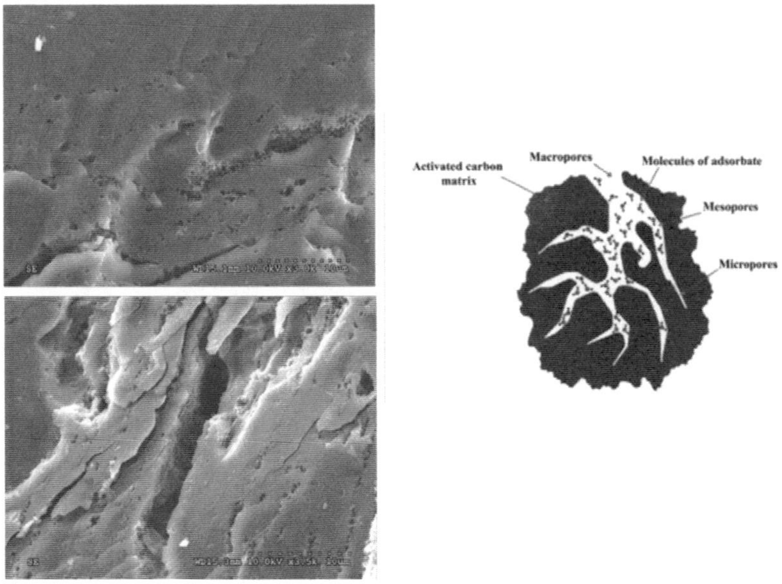

Figure 11 SEM for ABAC samples treated with NaOH (left) and scheme for the pore structure (right) [13]

The presence of these varied pore sizes contributes to the increased efficiency of CO_2 adsorption. Macropores provide the necessary pathways for the CO_2 molecules to penetrate the internal structure of the carbon, while meso- and micropores contribute to the actual surface area available for adsorption, capturing and retaining the CO_2 molecules.

4. Conclusions

In comprehensive CO_2 adsorption tests conducted across varying concentrations, anthracite-based activated carbon (ABAC) demonstrated robust efficacy, matching or surpassing that of commercial activated carbons in both Direct Air Capture (DAC) and Point Source Capture (PSC) configurations. Notably, when treated with NaOH, ABAC showcased a remarkable capacity to absorb twice the amount of CO_2 compared to its commercial counterparts, particularly at a concentration of 13%. This enhanced performance is attributed to the chemical modifications induced by the NaOH treatment, which significantly augment the adsorptive properties of ABAC. Furthermore, the durability of ABAC was evidenced by its consistent performance over multiple adsorption cycles in its reactivated form, indicating its potential for long-term application in industrial settings. Complementary analyses using Accelerated Surface Area and Porosity (ASAP) and Scanning Electron Microscopy (SEM) provided deeper insights into the material's structure, confirming that the highly porous nature of ABAC is a key factor in its superior CO_2 adsorption capacity. These findings collectively affirm the viability of ABAC as an effective, sustainable alternative for CO_2 capture technology, suggesting promising avenues for future enhancements and applications in environmental management strategies.

5. Acknowledgements

I extend my deepest gratitude to the Energy Research Center, particularly Carlos Romero and Zheng Yao, for their invaluable guidance and support throughout my research. Their expertise and insights have significantly shaped the direction and success of my work. I would also like to thank Blaschak Anthracite for their generous sponsorship of my research. Special thanks to Tom Lowe, Greg Driscoll, Boyd Kreglow, and Harold Schobert for their encouragement and substantial contributions to my projects. Furthermore, I am grateful for the support from the Pennsylvania Infrastructure Technology Alliance, which has played a crucial role in facilitating the resources needed for my research. Each of these individuals and organizations has had a profound impact on my studies, and I am immensely thankful for their collaboration and commitment.

5. References

[1] EPA Website, "Overview of Greenhouse Gases."

[2] P. FREUND and M. HAINES, "Precombustion Decarbonisation for Power Generation," *Greenhouse Gas Control Technologies - 6th International Conference*, pp. 1571–1574, Jan. 2003, doi: 10.1016/B978-008044276-1/50248-8.

[3] K. Jordal, M. Anheden, J. Yan, and L. Strömberg, "Oxyfuel combustion for coal-fired power generation with CO2 capture—Opportunities and challenges," *Greenhouse Gas Control Technologies*, pp. 201–209, Jan. 2005, doi: 10.1016/B978-008044704-9/50021-5.

[4] M. M. Hossain and H. I. de Lasa, "Chemical-looping combustion (CLC) for inherent CO2 separations—a review," *Chem Eng Sci*, vol. 63, no. 18, pp. 4433–4451, Sep. 2008, doi: 10.1016/J.CES.2008.05.028.

[5] J. Oexmann, C. Hensel, and A. Kather, "Post-combustion CO2-capture from coal-fired power plants: Preliminary evaluation of an integrated chemical absorption process with piperazine-promoted potassium carbonate," *International Journal of Greenhouse Gas Control*, vol. 2, no. 4, pp. 539–552, Oct. 2008, doi: 10.1016/J.IJGGC.2008.04.002.

[6] R. Stuart Haszeldine, "Carbon capture and storage: how green can black be?," 2009. doi: 10.1126/science.1172246.

[7] Y. Liang, D. P. Harrison, R. P. Gupta, D. A. Green, and W. J. McMichael, "Carbon dioxide capture using dry sodium-based sorbents," *Energy and Fuels*, vol. 18, no. 2, pp. 569–575, Mar. 2004, doi: 10.1021/ef030158f.

[8] C. Song, X. Xu, J. M. Andresen, B. G. Miller, and A. W. Scaroni, "Novel Nanoporous 'Molecular Basket' Adsorbent for CO2 Capture," *Stud Surf Sci Catal*, vol. 153, pp. 411–416, Jan. 2004, doi: 10.1016/S0167-2991(04)80288-X.

[9] B. Sadeghalvad, H. Ebrahimi, and A. Azadmehr, "CO2 capture by adsorption," *Emerging Carbon Capture Technologies: Towards a Sustainable Future*, pp. 63–89, Jan. 2022, doi: 10.1016/B978-0-323-89782-2.00012-0.

[10] G. Song *et al.*, "Anthracite coal-based activated carbon for elemental Hg adsorption in simulated flue gas: Preparation and evaluation," *Fuel*, vol. 275, p. 117921, Sep. 2020, doi: 10.1016/J.FUEL.2020.117921.

[11] Lattimer mammoth vein representative analysis - blaschak ..., https://www.blaschakanthracite.com/wp-content/uploads/Lattimer-Mammoth-Vein-Rep-Analysis.pdf (accessed Dec. 12, 2024).

[12] Y. Gao, Q. Yue, B. Gao, and A. Li, "Insight into activated carbon from different kinds of chemical activating agents: A review," *Science of The Total Environment*, vol. 746, p. 141094, Dec. 2020, doi: 10.1016/J.SCITOTENV.2020.141094.

[13] J. Serafin and B. Dziejarski, "Activated carbons—preparation, characterization and their application in CO2 capture: A review," *Environmental Science and Pollution Research*, vol. 31, no. 28, pp. 40008–40062, Jun. 2024, doi: 10.1007/s11356-023-28023-9.

[14] *BPL® 4x6 Granular Activated Carbon Data Sheet*. Calgon Carbon Corporation, 2015

[15] CO2 Meter, https://www.co2meter.com/collections/co2-sensors (accessed Dec. 12, 2024).

2024 International Pittsburgh Energy and Carbon Management Conference

October 7 - 9, 2024

Abstract Submission

PROGRAM TOPIC:
COMBUSTION TECHNOLOGIES

Impact of a Bluff Body on a Pressurized Oxy-fuel Flame

Presenting Author (Lei Li, GRA, West Virginia University, Morgantown, West Virginia 26506, USA)
Contact Information: ll00023@mix.wvu.edu, 304-435-5325

Co-Authors (V'yacheslav Akkerman, Professor, West Virginia University, Morgantown West Virginia 26506, USA)
Contact Information: Vyacheslav.Akkerman@mail.wvu.edu, 605-815-9021
Co-Author: Richard L. Axelbaum, Professor, Washington University in Saint Louis, Saint Louis, MO, USA
Contact Information: axelbaum@wustl.edu
Duarte Magalhaes Dr., Washington University in Saint Louis, Saint Louis, MO, USA
Contact Information: duartenuno@wustl.edu
Mao Cheng, Dr., Washington University in Saint Louis, Saint Louis, MO, USA
Contact Information: chengmao@wustl.edu

Abstract

Designing an effective burner is vital for the development of pressurized oxy-fuel combustion technologies. In the present work, turbulent jet diffusion burners are adopted for a pressurized oxy-combustor, with a bluff-body employed to create a recirculation zone, thereby stabilizing the flame in such a combustor. The objective of this numerical study is to perform a systematic analysis of the characteristics of such a pressurized non-premixed flame. Specifically, a 15-bar pressurized oxy-fuel combustor of power 100 kW_{th} is modeled by means of the Ansys FLUENT commercial platform, using the Reynolds-averaged Navier-Stokes (RANS) approach. The present work focuses on identifying the aerothermodynamic features of the pressurized oxy-fuel burner with a disk-shape bluff-body. It is shown that the fuel-to-oxidizer stream momentum ratio has a great impact on the temperature profile of the down-fired, co-axial, pressurized oxy-coal, diffusion flame. A parametric study of the blockage ratio of the burner identifies an optimal range for this pressurized burner.

Keywords: Pressurized oxy-combustion; Burner design; Reynolds-averaged Navier-Stokes (RANS) modelling; Carbon capture

1. Introduction

Pressurized oxy-combustion (POC) has become an emerging research direction in the field of green gas emission reduction, being an important application of carbon capture and storage [1,2]. Designing an effective burner is vital for the development of pressurized oxy-fuel combustion technologies, especially that with coal as a fuel – due to the complexity of coal combustion.

The oxy-fuel combustion reviews [3–5] highlighted the differences between air and oxy-combustion, such as the benefit of oxy-combustion, radiative heat transfer, flammable regimes, laminar burning velocity etc. Holtmeyer *et al.* [6] studied the impact of biomass-coal co-firing on the volatile flame length in atmospheric oxy-combustion in terms of the particle size and cofiring ratio. Karaismail *et al.* [7] investigated the impact of the stoichiometric mixture fraction on the radiative heat flux in POC by means of large-eddy simulations (LES). Gardner et al. [8] summarized the atmospheric oxy-combustion reaction with power ranging from 20 to 300 kWth, in which TU Darmstadt and RWTH Aachen University, Germany, employed the burner with features of annuli, bluff-body and swirl. In the U.S., the current existing pressurized oxy-coal burners are at Washington University in St. Louis (WUSTL) [9] and Brigham Young University [8]. The burner developed in this work is designed for the WUSTL pressurized oxy-combustion facility. Although swirling could enhance flame stability for oxy-fuel combustion, the centrifugal force created by the swirl flow throws the large fuel particles towards the furnace wall, leading to particle deposition. Therefore, no swirl is used in this design; instead, the recirculation mechanism to stabilize the POC flame depends on the design of the bluff-body. The relevant parameters of the burner design scrutinized in this work are the mass flow rate of the oxidizer and the size of bluff-body.

2. Methodology

2.1 Turbulence model

Among numerous existing turbulent models, the Reynolds-averaged Navier-Stokes (RANS) approach with the k-ω shear stress transport (SST) model is chosen for most simulations in this study. The continuity and momentum equations, respectively, read

$$\frac{\partial}{\partial x_i}(\rho u_i) = 0, \tag{1}$$

$$\frac{\partial}{\partial t}(\rho u_i) + \frac{\partial}{\partial x_j}(\rho u_i u_j) = -\frac{\partial p}{\partial x_i} + \frac{\partial}{\partial x_j}\left[\mu\left(\frac{\partial u_i}{\partial x_j} + \frac{\partial u_j}{\partial x_i} - \frac{2}{3}\delta_{ij}\frac{\partial u_l}{\partial x_l}\right)\right] + \frac{\partial}{\partial x_j}\left(-\rho\overline{u'_i u'_j}\right), \tag{2}$$

where $-\rho\overline{u'_i u'_j}$ is the Reynolds stress tensor, which has to be modeled in order to close Eq. (2). The k-ω SST model employs two more partial differential equations (PDEs) determining the specific turbulence kinetic energy (TKE), k, and the specific dissipation rate, ω, namely:

$$\frac{\partial}{\partial t}(\rho k) + \frac{\partial}{\partial x_i}(\rho k u_i) = \frac{\partial}{\partial x_j}\left(\Gamma_k \frac{\partial k}{\partial x_j}\right) + G_k - Y_k, \tag{3}$$

$$\frac{\partial}{\partial t}(\rho \omega) + \frac{\partial}{\partial x_i}(\rho \omega u_i) = \frac{\partial}{\partial x_j}\left(\Gamma_\omega \frac{\partial \omega}{\partial x_j}\right) + G_\omega - Y_\omega, \tag{4}$$

where the source terms G_k and G_ω describe the generation of k and ω, respectively, while

$$\Gamma_k = \mu + \frac{\mu_t}{\sigma_k}, \tag{5}$$

$$\Gamma_\omega = \mu + \frac{\mu_t}{\sigma_\omega} \tag{6}$$

represent the effective diffusivity of k and ω. The terms Y_k and Y_ω describe the dissipation of k and ω, while σ_k and σ_ω stand for the turbulent Prandtl numbers for k and ω, respectively.

2.2 Combustion model

Turbulent reacting flow could be predicted by the turbulence-chemistry interaction model [10]. Multiple simultaneous chemical reactions such as volumetric reactions and particle surface reactions are considered in the present study. The quantity of each species is given by the species transport equation, associated with convection, diffusion and reaction,

$$\frac{\partial}{\partial t}(\rho Y_i) + \nabla\cdot(\rho\vec{v}Y_i) = -\nabla\cdot\vec{J_i} + R_i + S_i, \tag{7}$$

where ρ is the density, $\vec{v}$ is the velocity vector, Y_i represents the local mass fraction of ith species, R_i is the net rate of production of species i by chemical reaction, and S_i is the rate of creation by addition from the dispersed phase plus any user-defined sources. The mass diffusion of turbulent flow is given by

$$\vec{J_i} = -\left(\rho D_{i,m} + \frac{\mu_t}{Sc_t}\right)\nabla Y_i - D_{T,i}\frac{\nabla T}{T}, \tag{8}$$

where $\vec{J_i}$ is the diffusion flux and $D_{i,m}$ is the mass diffusion coefficient for species i, $D_{T,i}$ is the thermal diffusion coefficient, and $Sc_t = \mu_t/\rho D_t$ is the turbulent Schmidt number, with the turbulent viscosity μ_t and D_t is the turbulent diffusivity.

2.3 Particle properties

The fuel used in the simulation is Powder River Basin (PRB) coal, which properties are listed in Table 1.

Table 1: The Powder River Basin (PRB) coal properties.

Fuel	Proximate, wt. % ar				Ultimate, wt. % daf				HHV J/kg
	Fixed carbon	Volatile matter	Ash	Moisture	C	H	N	S	
PRB	44	40	8.3	7.3	74.7	5.5	1.6	1.2	2.54e+07

The particles under consideration are assumed to be spherical, with the particle size distribution obeying the Rosin-Rammler (RR) law [10]

$$Y_d = e^{-(d/\bar{d})^n}, \tag{9}$$

where Y_d designates the mass fraction of the particles, which diameter exceeds d, $\bar{d}$ is the mean diameter, and n is the size distribution parameter, 3.5. The probability density function (PDF) can be obtained by differentiating Eq. (9). The mean diameter of this particulate cloud is about 65 μm and the particle diameter ranges from 5 to 200 μm.

3. Results and Discussion

3.1 Isothermal Analysis of Confined Flow over a Bluff-body

To analyze the features of the recirculation zone induced by a bluff-body, an annular oxidizer flow without central coal carrier is considered first and the velocity of this oxidizer flow, U_e, is shown in Figure 1. The velocities of the oxidizer stream (0.122, 0.305, 0.488, 0.61 m/s) are corresponding to consume coal thermal input (20, 50, 80, 100 kW). The stoichiometric ratio (SR), λ, (the ratio of the actual oxygen mass over the mass of oxygen required for stoichiometric combustion) used to calculate the oxidizer flow rate here is 1.054.

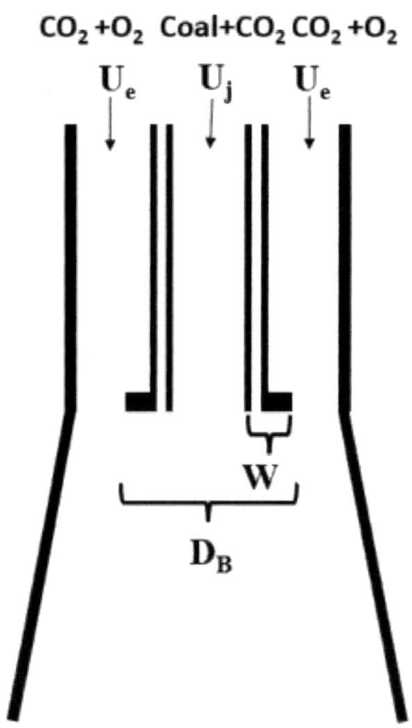

Figure 1: A schematic diagram for the pressurized oxy-combustion (POC) burner, D_B indicates the outer diameter of the disk, W the width of the base of the bluff-body.

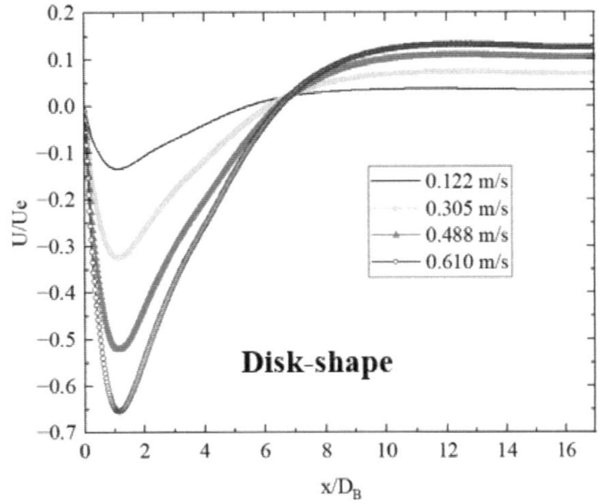

Figure 2: Centerline evolution of the non-dimensional mean axial velocity (U/U_e) for the disk-shape bluff-body and various velocities.

Table 2: A parametric study of the stoichiometric ratio (SR) λ.

Stoichiometric ratio λ	**Oxidizer mass flow rate (kg/s)**	**Oxidizer flow velocity U_e (m/s)**	**Coal carrier velocity U_j (m/s)**
0.8	0.025	0.46	0.21
1.054	0.033	0.61	
1.2	0.038	0.69	
1.4	0.047	0.87	

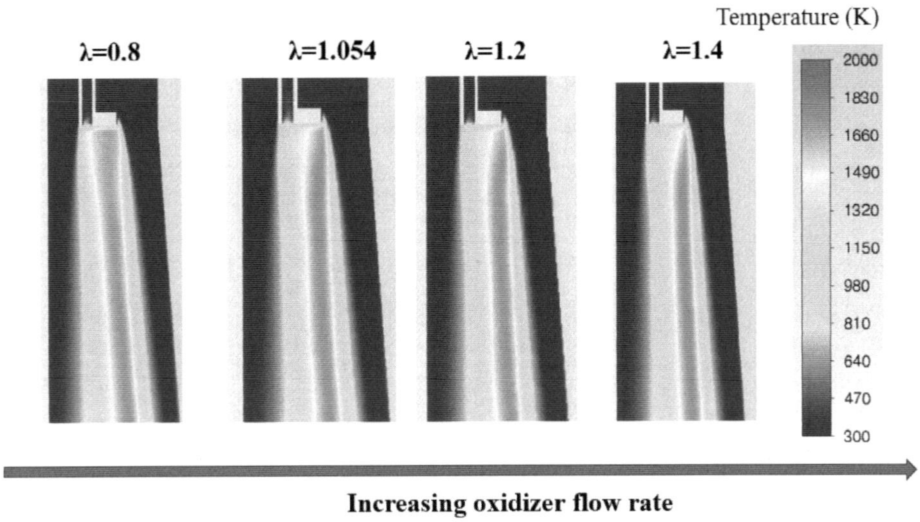

Figure 3: Temperature contours for various oxidizer flow rates.

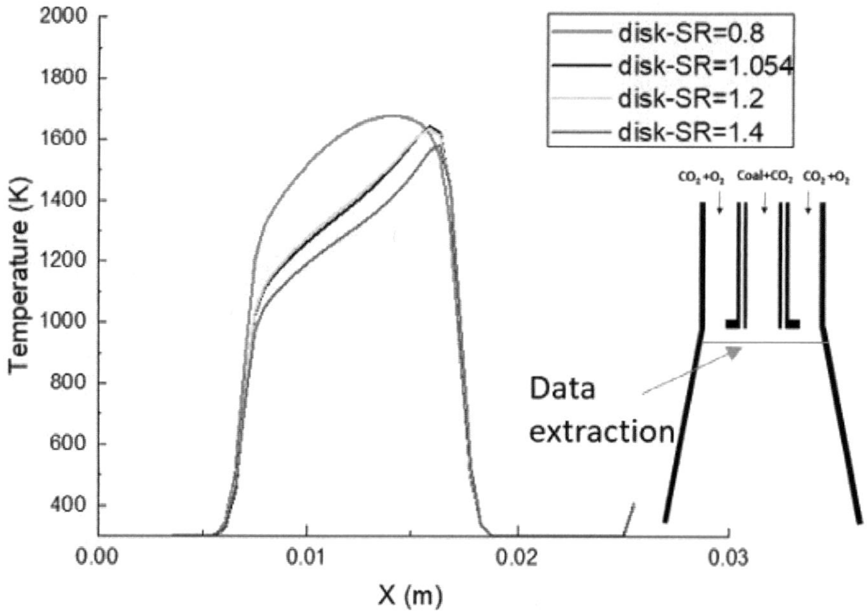

Figure 4: Temperature profiles in the near-burner region (2 mm away) of the bluff-body.

The centerline evolution of the isothermal non-dimensional mean axial velocity (U/U_e) for the disk bluff-body is extracted and depicted in Figure 2, in which U_e stands for the mean axial annular velocity of the oxidizer-flow. It is observed that the recirculation velocity increases with the velocity of the oxidizer flow. The length of the recirculation zone of the disk-shape burner changes from $5.5D_B$ to $6.7D_B$.

First, the impact of the flow rate of the oxidizer stream on the flame is investigated for disk-shape burner. The total thermal input from coal for all cases is 100 kW. The stoichiometric ratio (SR) λ (the ratio of the actual oxygen mass over the mass of oxygen required for stoichiometric combustion) is used to indicate the change of the oxidizer flow rate. The oxidizer mass flow rates as well as their respective mean velocities are listed in Table 2. Figure 3 shows how the flame mode changes with the oxidizer flow rate for the disk-shape burner. The high temperature zone of the flame in the near-burner region decreases with the SR.

Figure 4 presents the radial temperature profiles in the near-burner region (2 mm away from the burner tip). The peak temperature drops when the SR increases. Especially, when SR = 0.8, the hot area in the near-burner region grows significantly. Since the velocity of the oxidizer stream greatly exceeds that of the fuel stream, the recirculation is mainly produced by the oxidizer stream. The recirculation flow rate increases with the SR and diminishes the temperature in the vicinity of the burner.

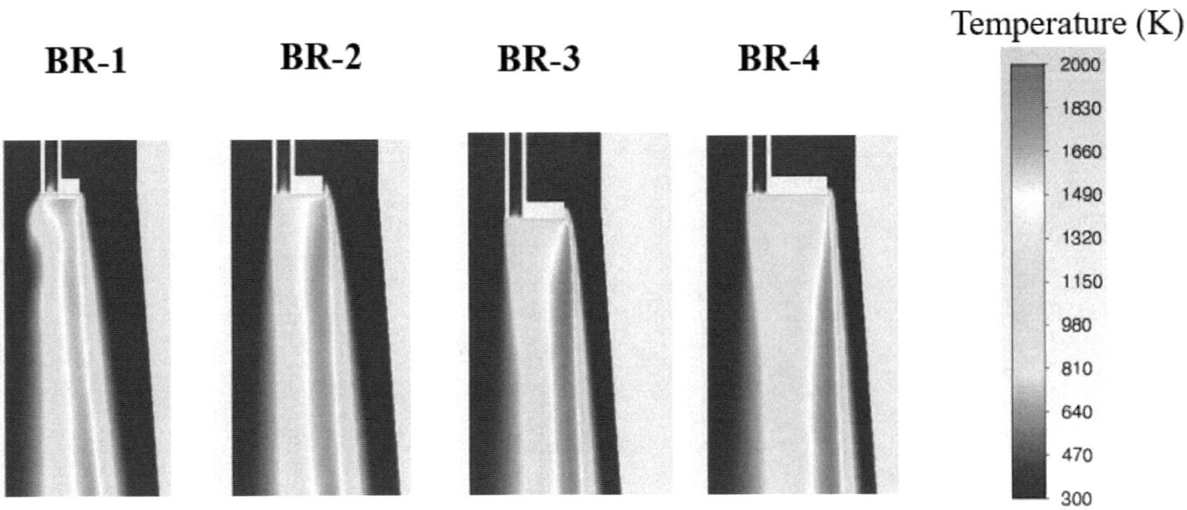

Figure 5: The colorful temperature snapshots in the near-burner region of the disk-shape burner of various blockage ratios (BR).

The blockage ratio (BR) is another important parameter for this burner, which is associated with the recirculation velocity of oxidizer stream. Figure 5 shows how the blockage ratio influences the flame; the BR-1 is a threshold value to sustain the POC flame, since the flame was not viable at lower BRs in the simulation. The flame becomes wrinkled since the burner tip surface is not large enough to accommodate the flame. The area of the high temperature zone decreases with the BR, thereby increasing the tendency of the flame detachment from the burner. It is interesting to note that the flow velocities in the recirculation zone are similar for different blockage ratios, because the increased recirculation flow rate could be accommodated by the increased recirculation zone, as shown in Figure 6.

A higher BR leads to a higher heat release from the reaction in the vicinity of the burner, as shown in Figure 7 and higher volatile concentration as seen in Figure 8.

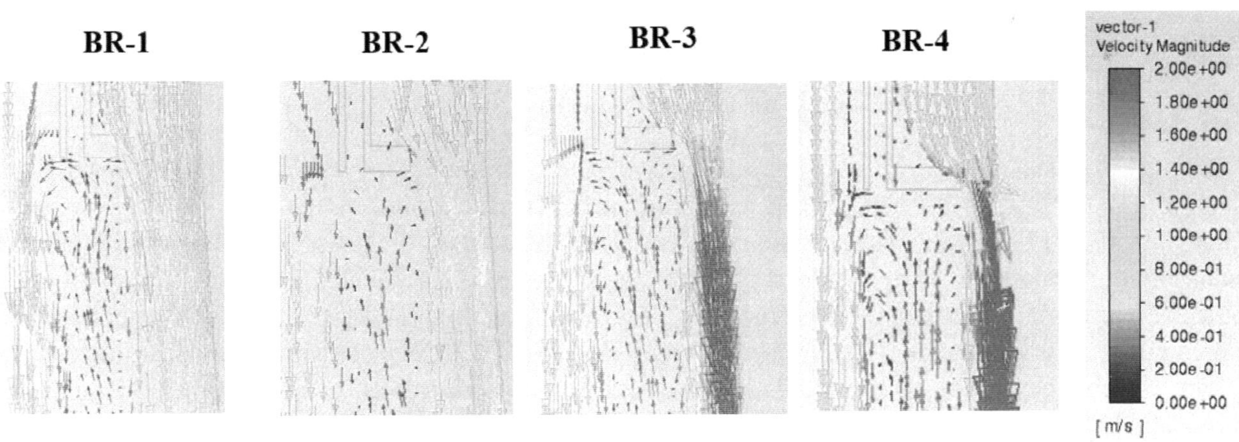

Figure 6: Velocity vector profiles in the near-burner region of the disk-shape burner of various BRs.

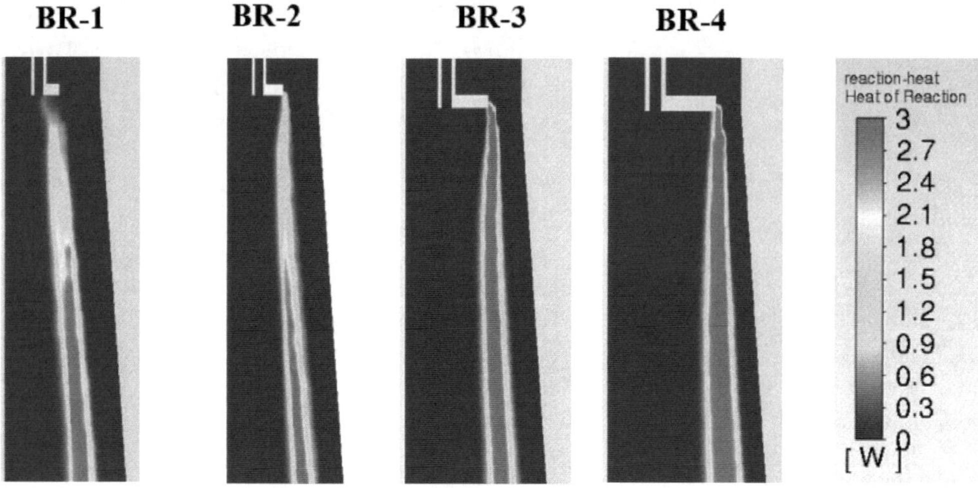

Figure 7: Heat of reaction in the near-burner region of the disk-shape burner of various BRs.

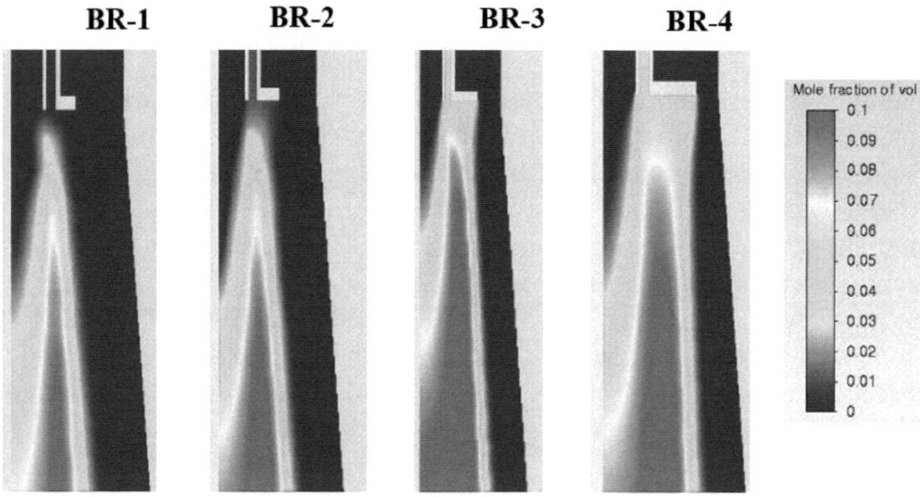

Figure 8: Volatile concentrations in the near-burner region of the disk-shape burner of various BRs.

4. Conclusions

This work focuses on the impact of a bluff-body on the stabilization of a pressurized oxy-coal flame. An isothermal analysis has been performed to identify the recirculation features of this burner. A parametric study on the oxidizer flow rate and size of the disk-shape bluff-body has been undertaken to ascertain the impact on the flame. Both parameters could lead to different flame modes in the vicinity of the burner. The isothermal oxidizer flow induces a recirculation flow, and the length of the recirculation zone changes only a little as the flow rate increases by a factor of 5. An increase in the stoichiometric ratio has a similar impact to that of an increase in the disk size, because both effects promote the oxidizer velocity. The lower threshold value of BR to obtain a stabilized flame for the disk-shape burner is BR-1.

5. Acknowledgements

This material is based upon works supported by the Department of Energy, USA, under Award Number DE-FE0031925. The views and opinions of authors expressed herein do not necessarily state or reflect those of the United States Government or any agency thereof. This work was also supported by the Consortium for Clean Coal Utilization (CCCU) at Washington University in St. Louis.

6. References

[1] Gopan A, Verma P, Yang Z, Axelbaum RL. Quantitative analysis of the impact of flue gas

recirculation on the efficiency of oxy-coal power plants. Int J Greenh Gas Control 2020;95:102936. https://doi.org/10.1016/j.ijggc.2019.102936.
[2] Xia F, Yang Z, Adeosun A, Gopan A, Kumfer BM, Axelbaum RL. Pressurized oxy-combustion with low flue gas recycle: Computational fluid dynamic simulations of radiant boilers. Fuel 2016;181:1170–8. https://doi.org/10.1016/j.fuel.2016.04.023.
[3] Buhre BJP, Elliott LK, Sheng CD, Gupta RP, Wall TF. Oxy-fuel combustion technology for coal-fired power generation. Prog Energy Combust Sci 2005;31:283–307. https://doi.org/10.1016/j.pecs.2005.07.001.
[4] Wall T, Liu Y, Spero C, Elliott L, Khare S, Rathnam R, et al. An overview on oxyfuel coal combustion-State of the art research and technology development. Chem Eng Res Des 2009;87:1003–16. https://doi.org/10.1016/j.cherd.2009.02.005.
[5] Chen L, Yong SZ, Ghoniem AF. Oxy-fuel combustion of pulverized coal: Characterization, fundamentals, stabilization and CFD modeling. Prog Energy Combust Sci 2012;38:156–214. https://doi.org/10.1016/j.pecs.2011.09.003.
[6] Holtmeyer ML, Li G, Kumfer BM, Li S, Axelbaum RL. The impact of biomass cofiring on volatile flame length. Energy and Fuels 2013;27:7762–71. https://doi.org/10.1021/ef4013505.
[7] Karaismail FN, Gopan A, Axelbaum RL, Celik I, Kumfer BM. Large eddy simulations of staged pressurized oxy-combustion. 2018 Spring Tech Meet East States Sect Combust Institute, ESSCI 2018 2018;2018-March:1–8.
[8] Gardner S, Carpenter C, Tuia J, Skousen A, Badger D, Thomas S, et al. Design and performance of a laboratory scale pressurized oxy-coal reactor. Results Eng 2023;18:101192. https://doi.org/10.1016/j.rineng.2023.101192.
[9] Gopan A. Studies in Pressurized Oxy-Combustion : Process Development and Control of Radiative Heat Transfer. Washington University in St. Louis, 2017.
[10] Ansys inc. ANSYS Fluent Theory Guide. ANSYS Inc, USA 2022.

2024 International Pittsburgh Coal Conference
October 7 - 9, 2024
Abstract Submission
PROGRAM TOPIC:
CLEAN DEMONSTRATION AND COMMERCIAL PROJECTS
Exact Title of Paper:

PLANNED UNIT WILL PRODUCE REE ALLOYS FOR PERMANENT MAGNETS

Howard McClintic, Director of REEs & Recycling, Concurrent Technologies Corporation, Washington, DC, U.S.A.
McClintH@ctc.com, (202) 689-4586

Abstract

Concurrent Technologies Corporation (CTC) has received funding to design and is constructing a demonstration-sized electric arc furnace (EAF) that will achieve multiple high temperature gasification goals. The high temperature furnace is superior to those that Hitachi Metals has sold and that operate in China, because CTC's furnace achieves higher temperatures (Hitachi 1900 degrees centigrade and CTC's 3400 degrees centigrade). See our attached 2021 IPCC Paper. If the demonstration unit is successful, the demonstration and commercial units would complete the long sought domestic supply chain: the United States would be in a position to produce alloys or metallic neodymium iron boron (NdFeB) permanent magnets. CTC has the capability to 3-D print these metallic magnets.

Critical Smelting temperatures for REE's

Name	Symbol	No.	Melting	% Import for Magnets
Neodymium	Nd	60	1022°C	30%
Boron	B	5	2077°C	1%
Iron	Fe	26	1600°C	69%

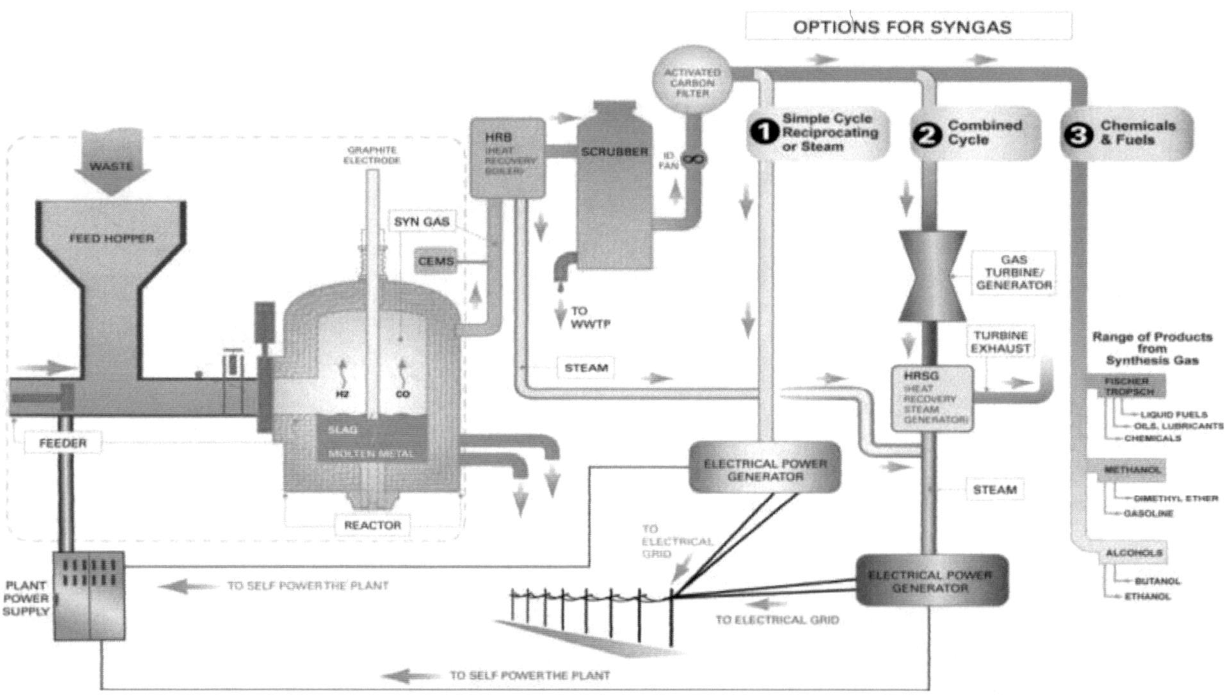

EXECUTIVE SUMMARY

The US Congress and various Presidents have sought to reduce the Nation's reliance on China for rare earth elements (REEs), since there is hardly any minable, geologically-based resources (with the exception of Mountain Pass, CA and (radioactive) monazites in placer sands of Georgia and Florida) and no refining capability to produce REE-based (NdFeB) permanent magnets, the "gap" to fill was refining of coal fly ash that has a low concentration of REEs (~500 parts per million [ppm]).

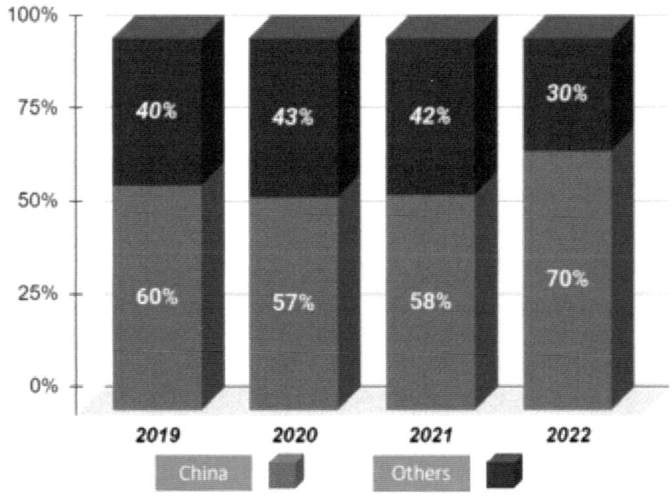

Global Mining Production for All Rare Earth Minerals (REMs)
Source: (USGS, 2021, 2022, 2023)

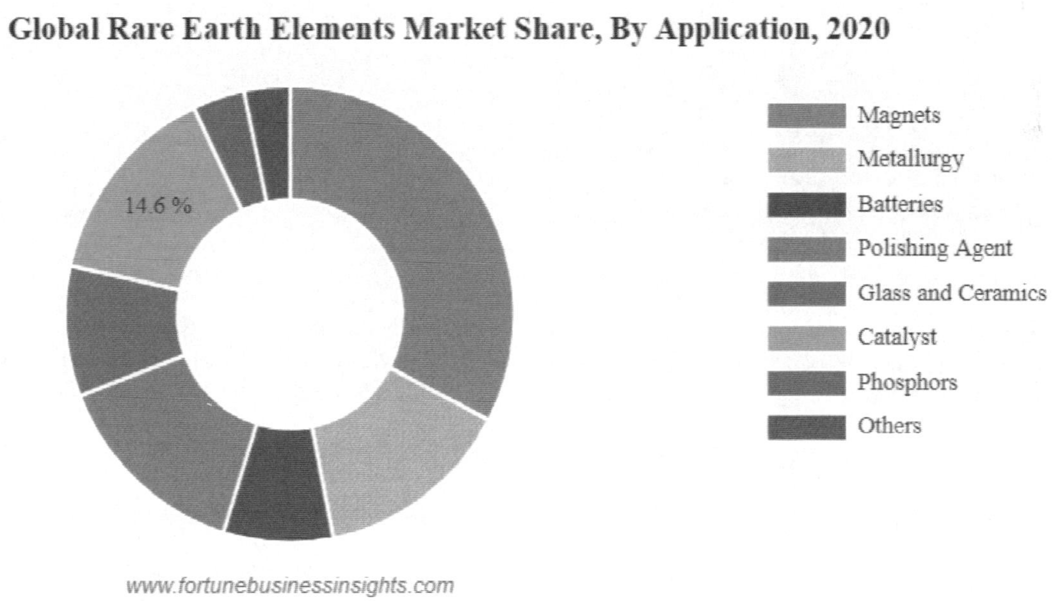

CTC's "sweet spot" is alloying, by gasification, the REOs, ultimately into permanent magnets. The four REE metals for permanent magnets for EVs and wind energy are: Nd, Pr, Tb and Dy. CTC's demonstration furnace will be the only smelting technology – hence, it is an essential link to completing the domestic supply chain.

BACKGROUND

After the demonstration furnace proves to be successful, CTC will be refining into permanent magnet alloys are 30% neodymium, 69% iron and 1.0% boron. CTC's furnace will achieve **some 3,000 degrees centigrade**, which is more than adequate to produce alloys for NdFeB magnets: 1,022°C for neodymium, 30% of the magnet; 1,538°C for iron, 69% of the magnet; and, 2,076°C for smelting boron, 1% of the magnet.

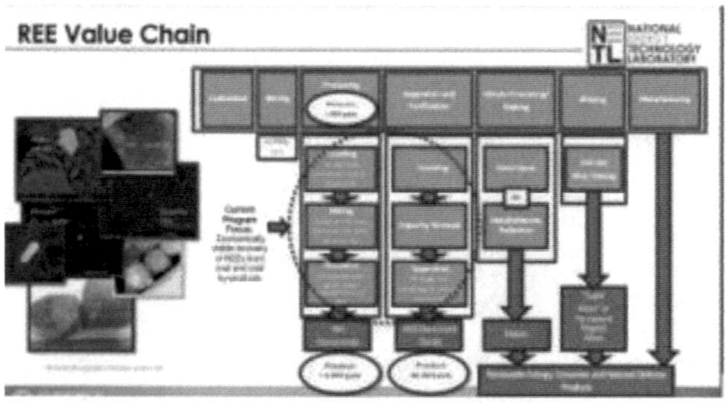

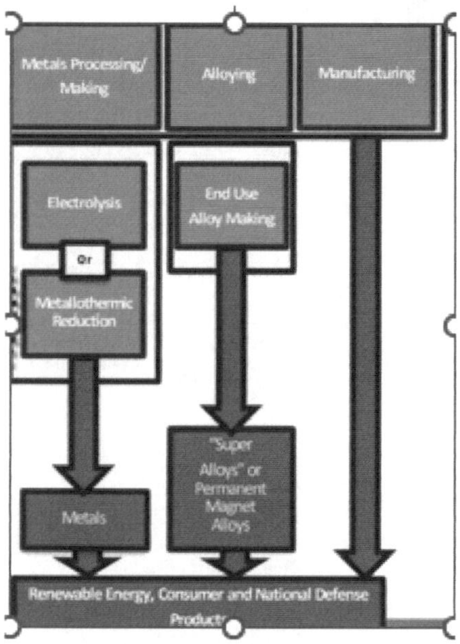

Production at both the mining and refining stages is controlled by quotas assigned to **six state-owned Chinese enterprises**, which are highly integrated throughout the rare earth supply chain. As trading relations between these two trading partners have become progressively more strained, a US/domestic supply chain has been deemed in the "national interest".

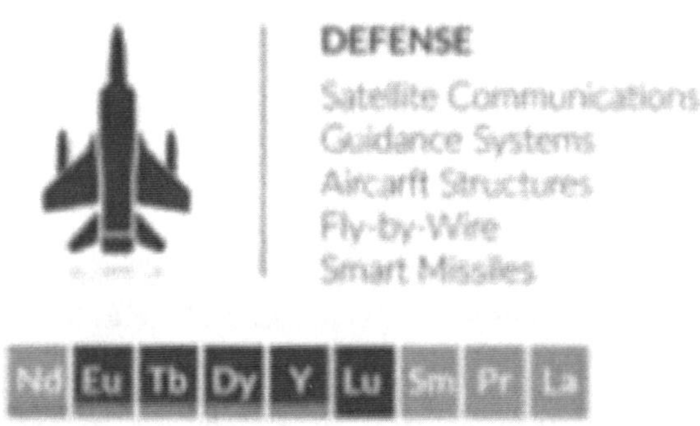

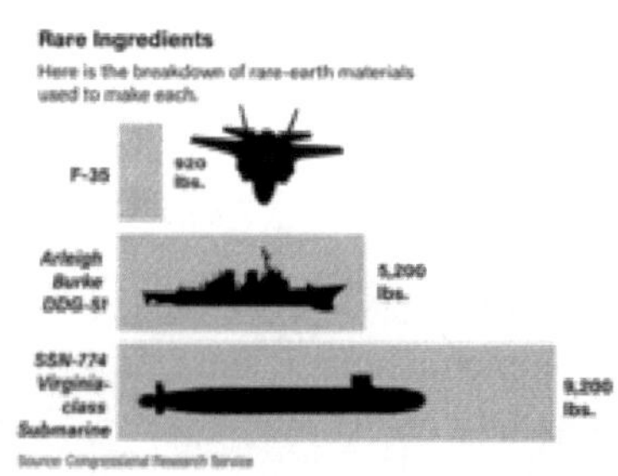

EXPERIMENTAL WORK

After the demonstration unit is constructed, operational, and everyone is convinced of its operability for extracting REEs from REOs (and gasifying single-use plastic bags) funding for a FEED (front-end engineering and design) study will be needed. **FEED studies are site specific, not generic**. CTC will work with Robindale Energy's senior management to develop a commercial plant, at one of the three western Pennsylvania waste coal plants it supports, perhaps at the Seward Plant in New Florence, PA https://robindale.energy/companies/seward-generation/).

These temperatures will be sufficient to gasify samarium (1072°C) as well as cobalt (1495°C) for almost comparable permanent magnets. Samarium is derived from monazite sands in Georgia and Florida.

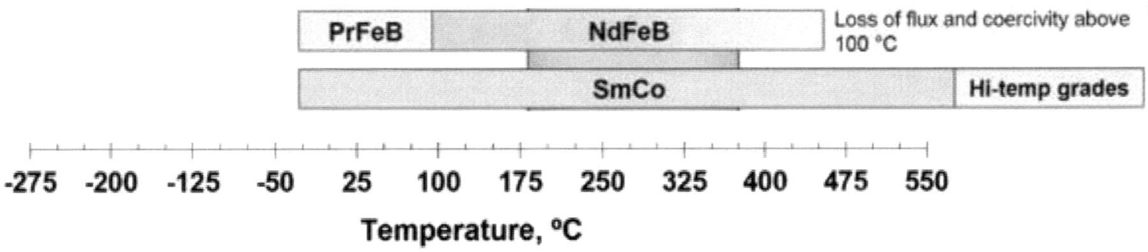

Neodymium magnets are sensitive to high temperatures, and their performance can degrade when exposed to elevated heat. The typical maximum operating temperature for standard neodymium magnets is around 80 to 150 degrees Celsius, depending on the specific grade. Neodymium magnets are renowned for their unparalleled magnetic strength and are currently the strongest commercially available permanent magnets.

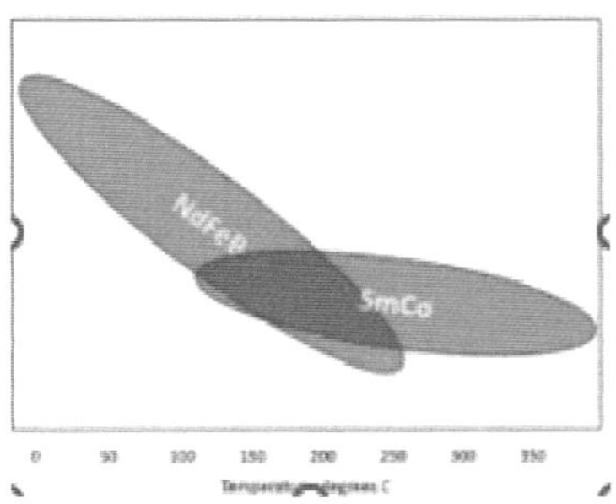

Samarium cobalt magnets exhibit better temperature stability. They can withstand higher temperatures without significant loss of magnetic properties. The working temperature range for samarium cobalt magnets is typically between 250°C and 350°C. Samarium cobalt magnets are not as strong as neodymium magnets, samarium cobalt magnets, but still possess high magnetic strength, making them suitable for applications where temperature stability is critical.

SUMMARY

Having appropriate support, CTC wants its planned high temperature, electric arc furnace (EAF) demonstration unit to be successful, which includes producing both rare earth element-based permanent magnets: neodymium and samarium (NdFeB and SmCo) ones. After that "proof of concept", the next steps are known and expensive.

A front-end engineering design (FEED) is undertaken by a qualified engineering-construction Firm; with some funding from the US Energy Department on a related Project, potential candidates were met. A FEED Study is exclusive to a specific and unique site; the possible Partner list one of the three western Pennsylvania waste coal plants that Robindale Energy supports, perhaps at the Seward Plant in New Florence, PA https://robindale.energy/companies/seward-generation/). Readers of our prior Papers know that REEs are not a new subject. Another Paper on the Profitability of Slag, by the same author, cites the same location in western Pennsylvania.

The construction of a demonstration unit to process REEs is a clear indication that CTC has the technology and is committed to providing a domestic supply chain for REEs, reducing the reliance on unreliable Chinese sources.

# # # #

Understanding the Regional Variations in Pay Thickness of Unconventional Reservoirs in the Powder River Basin

Prepared for:
AAPG RMS Meeting 2024

Prepared By:
Luke Fritz – Advanced Resources International, Inc.

October 7, 2024

Disclaimer

This report was prepared as an account of work sponsored by an agency of the United States Government. Neither the United States Government nor any agency thereof, nor any of their employees, makes any warranty, express of implied, or assumes any legal liability or responsibility for the accuracy, completeness, or usefulness of any information, apparatus, product, or process disclosed, or represents that its use would not infringe privately owned rights. Reference herein to any specific commercial product, process, or service by trade name, trademark, manufacturer, or otherwise does not necessarily constitute or imply its endorsement, recommendation, or favoring by the United States Government or any agency thereof. The views and opinions of authors expressed herein do not necessarily state or reflect those of the United States Government or any agency thereof.

NATIONAL ENERGY TECHNOLOGY LABORATORY

Advanced Resources International, Inc.

Agenda

- Introduction of the DOE-funded project: *Unlocking the Tight Oil Reservoirs of the Powder River Basin, Wyoming*
- Introduction of the Tight Oil Reservoirs in the Powder River Basin (PRB)
- Understanding the Turner Sandstone Pay
- Understanding the Mowry Shale Pay and where "Sweet Spots" may exist
- Conclusions/Summary
- Future Work

3

Advanced Resources International, Inc.

Introduction of the DOE-funded project: *Unlocking the Tight Oil Reservoirs of the Powder River Basin, Wyoming*

DE-FE0031779

Project Overview - Project Team

EORI

Project Participants

Industry Partner – Oxy

Principal Investigator – Enhanced Oil Recovery Institute (EORI)

Subaward – Advanced Resources International (ARI)

Contractor – NSI

Contractor – SIU

Contractor – PiriTech

University of Wyoming School of Energy Resources

OXY

Enhanced Oil Recovery Institute

Advanced Resources International, Inc.

SIU Carbondale

PIRI TECHNOLOGIES, LLC

Project Objectives

- Develop "optimized" completion design for emerging shale play
  - Characterize (basin-wide) **Mowry**, **Turner**, and **Niobrara** formations
  - Detailed geological mapping across PRB from well logs
  - Learn from recent and current completion practices
- Characterize reservoir rock
  - Collect and characterize core from pilot holes
  - Pore-scale visualization of fluid movement
  - Geomechanical studies: mineralogy, stress, permeability, fracture mechanics
- Drill and complete new horizontal well (field laboratory)
  - Incorporate latest technology to monitor completion and production in real time
- Refine completion design as informed by feedback from field laboratory
- Create development strategy with stakeholders
  - Developmental benefits, impacts, and challenges
  - Technical and economic viability for proposed strategy
  - Project risks and mitigation strategies

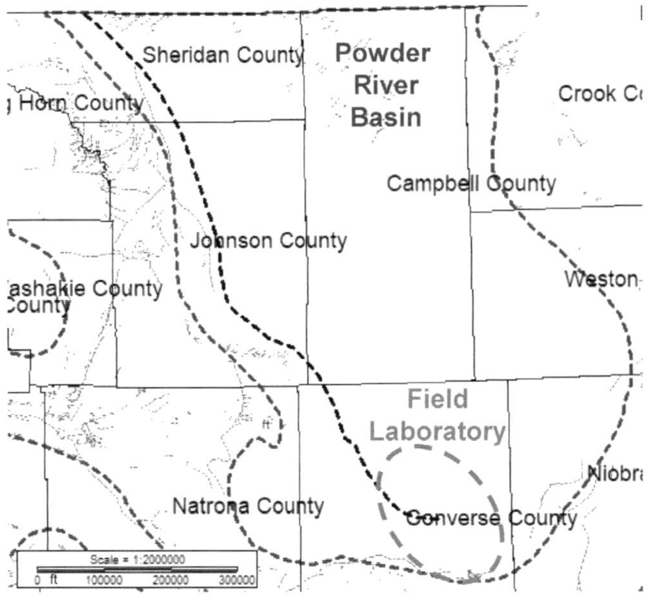

*Current Stage of Project

Advanced Resources International, Inc.

Introduction of the Tight Oil Reservoirs in the PRB

Heger, 2017

Introduction of the Tight Oil Reservoirs in the PRB

- Located in eastern Wyoming and Montana
- Succession of Paleozoic through Tertiary sedimentary rocks that overlie the Precambrian Basement
- Mowry Shale and Niobrara Formation are source rocks with increasing thermal maturity toward basinal axis

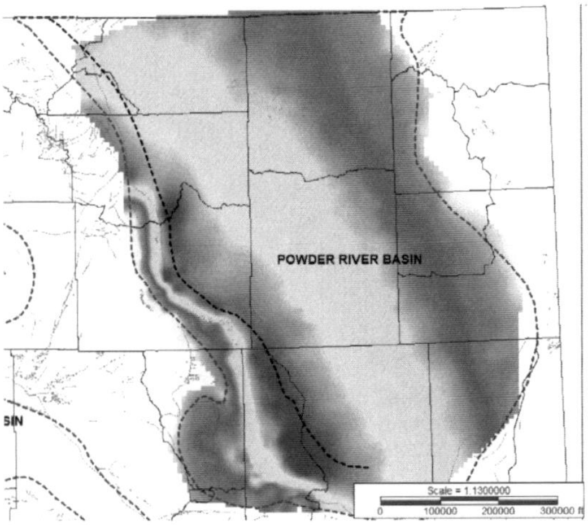

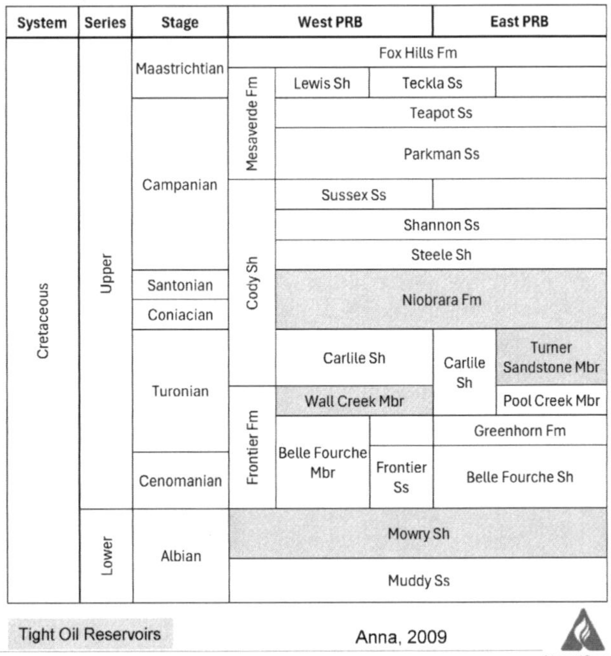

Anna, 2009

Mowry Shale

- Organic-rich mudstone deposited during **a major** marine transgression in the Cretaceous period
- 120 -> 400 ft thick throughout PRB
- During deposition there were fluctuations in sediment input and oxygenation levels. Persistent dysoxic to anoxic bottom-water conditions resulted in the preservation of organic matter
- Bentonite interbedded throughout, shown by high gamma ray spikes
- Subdivided into the Lower, Middle, and Upper members.
  - Middle Member has highest TOC and oil saturation
  - The maximum flooding surface also occurs in the Middle Member

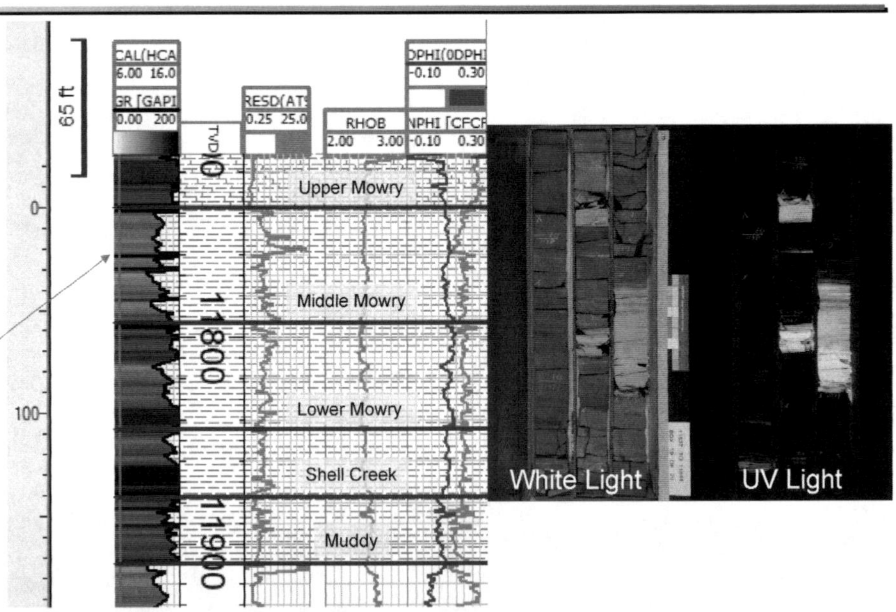

From the Field Laboratory in the Southern PRB

Turner Sandstone Member – Carlile Shale

- Very fine-grained, well-sorted sandstones interbedded with shales deposited during the Cretaceous
  - Varying degrees of bioturbation and mud content
  - The sandstone packages are often interbedded with mudstones or highly heterolithic mixtures of sandy mud
- Distal lower shoreface depositional environment
- Consists of three to four coarsening upward sandstone cycles that were deposited during an overall period of base level rise (Sonnenberg, 2020)
- Range in thickness from 150 feet to 200 ft across the southern PRB

Field Laboratory Thin Section | Field Laboratory Core

→ Intergranular pores

Advanced Resources International, Inc.

Completed Tasks Related to Geologic Modeling

- Uploaded field laboratory and public core data for the Mowry Shale and Turner Sandstone into geologic modeling software
- Picked tops for wells throughout the PRB to create structure, isopach, and pay maps
  - 2,099 Niobrara Formation tops
  - 2,244 Turner Sandstone tops
  - 1,118 Mowry Shale tops
  - Structure and isopach maps align with the geometry of previously published literature, confirming the quality and robustness of the generated dataset
- Created pay isopach maps for the Turner Sandstone
- Compared reservoir attributes to production metrics throughout the PRB

Turner Top Structure Map

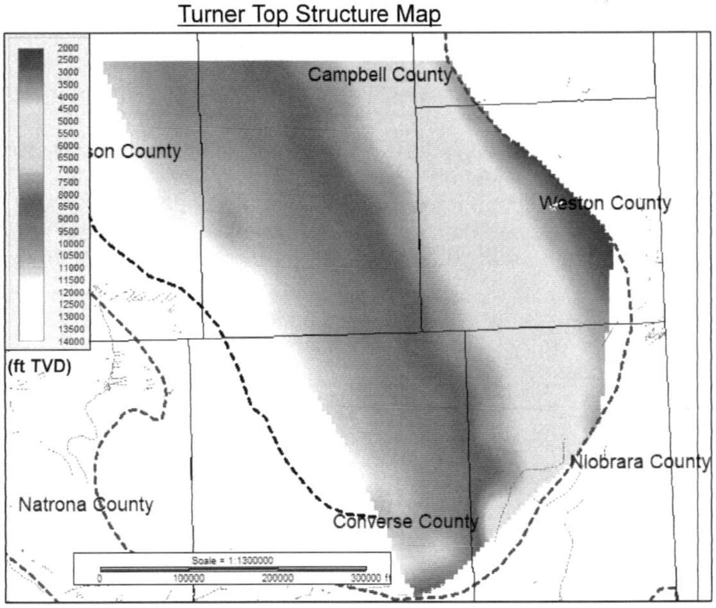

Advanced Resources International, Inc.

Understanding the Turner Sandstone Pay

Assigning Facies to the Turner Sandstone

- Compared assigned facies at the hand sample scale from Field Laboratory core with measured data such as oil saturation (S_o) and total porosity measurements
  - Clean Sandstone Facies indicates the pay zone in the Turner Sandstone
    - Clean Sandstone Facies is more coarse-grained and has elevated porosity and S_o
  - Investigated the petrophysical measurements (GR, RHOB) of the assigned facies. Unique, low GR/RHOB values for Clean Sandstone facies allowed for further petrophysical analysis across the PRB

	Facies #	Facies Name	Average Bioturbation Level	Frequency (ft)
	1.1	Massive to faintly bedded mudstone	1.6	7
	1.2	Sandy/Silty Mudstone	1.9	16
	1.3	Calcareous Sandy/Silty Mudstone (Limestone)	3.7	16
	1.4	Calcareous Laminated Mudstone, Siltstone, Sandstone (Limestone)	1.9	8
	2	Laminated Mudstone, Siltstone, Sandstone	1.7	20
	3.1	Bioturbated Heterolithic Silty/Sandy Mudstone	2.8	27
	3.2	Burrow Mottled Heterolithic Mudstone	3.7	7
	3.3	Bioturbated Heterolithic Calcareous Silty/Sandy Mudstone (Limestone)	3.1	14
	4.1	Bioturbated Argillaceous Sandstone	3.6	8
	4.2	Burrow Mottled Argillaceous Sandstone	3.8	70
Pay	4.3	Clean Sandstone	3.4	14

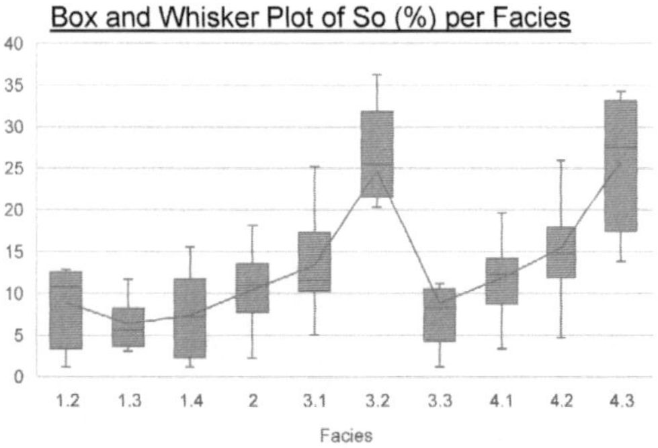

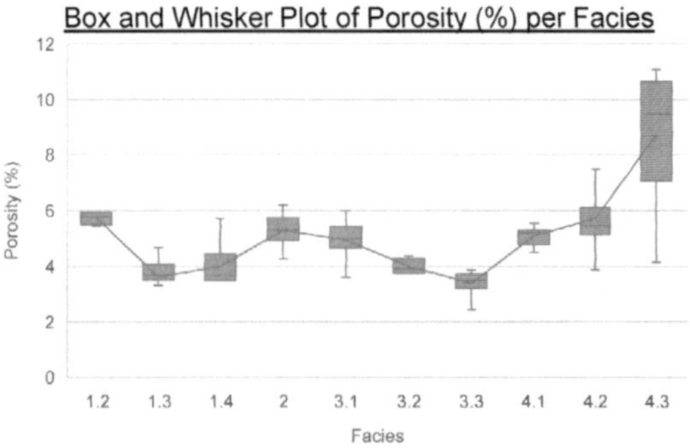

——— Mean

Advanced Resources International, Inc.

Turner Sandstone Pay Thickness

Field Laboratory Well

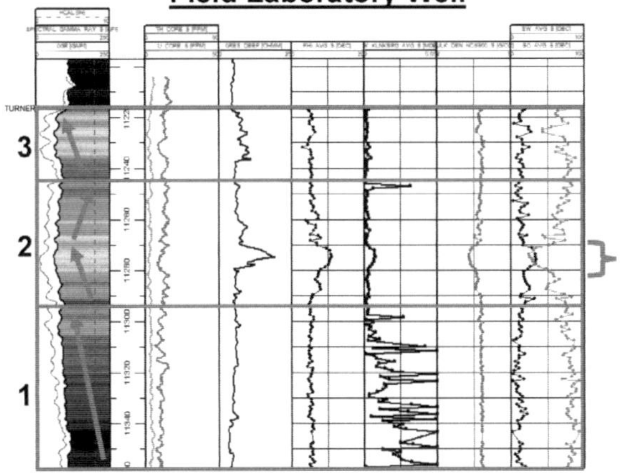

Pay Thickness Map

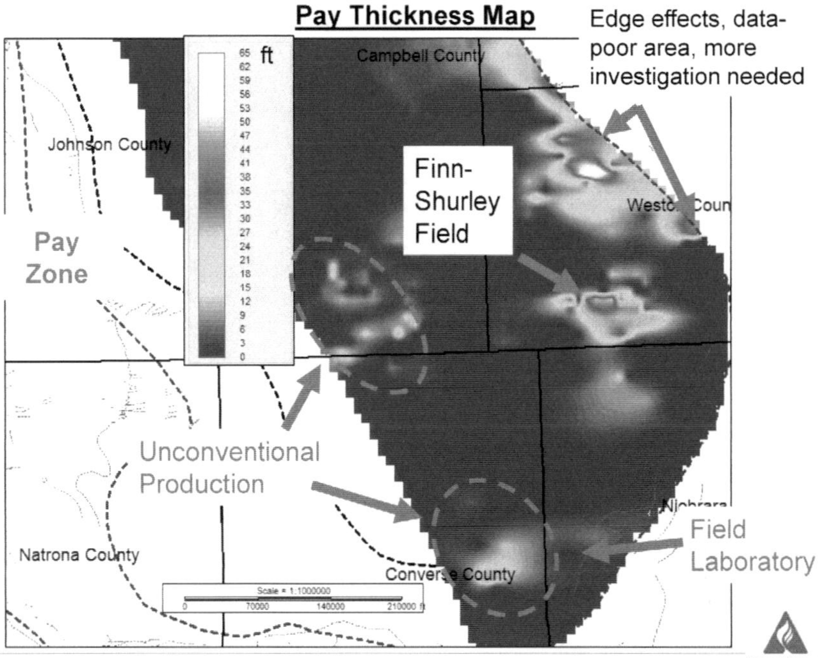

- Parameters used for Pay: GR <100, RHOB <2.5, CAL < 9, SWS <0.7; 384 pay zones calculated in the Turner
- SWS = Simandoux Water Saturation Equation
  - A=0.62, M = 2.15, N=2 (Milar, 2021)

13

Advanced Resources International, Inc.

Understanding the Mowry Shale Pay

Completed Tasks Related to the Field Laboratory's Mowry Core Data

- Correlations between facies groups originally classified by Purvis, et al. (2017) and core derived properties did not manifest within the Field Laboratory's data
  - Overall, at the hand sample scale (whole core), the Mowry Formation is largely a homogenous organic shale
  - Facies groups are not sensitive to the rock properties that drive well performance (TOC, oil saturation, porosity, etc)
  - Across the entire Mowry formation, TOC values range from 1.15% to 4.92%, with the highest average TOC occurring in the Middle Member

Mowry Facies

Facies #	Facies Name	Average Bioturbation Level	Frequency (ft)
1	Subtidal sheet, oxic	0	0
2	Subtidal sheet, suboxic	2.3	3
3	Subtidal sheet, dysoxic	1.3	28
4	Subital sheet, anoxic	0.3	111

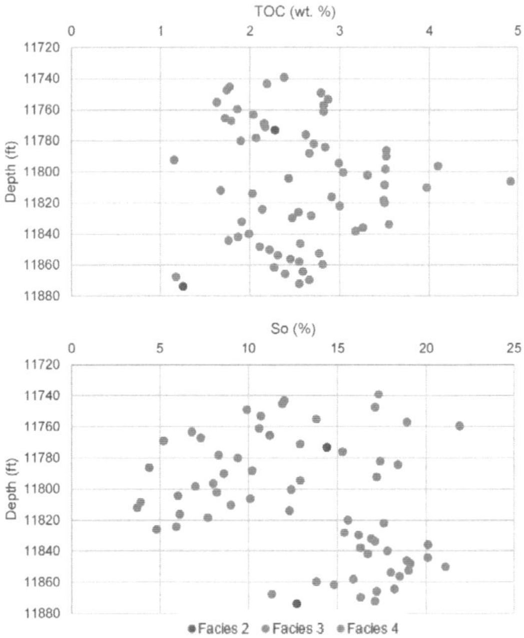

Assessing the Mowry Sweet Spots - Tmax (℃) of the Mowry Shale

- Hybrid Type II and Type III kerogen
  - More Type III to the West, closer to the terrestrial sediment source
- Three areas of higher maturity in the PRB
  - Similar trend seen in Finley (2021)
- 102 wells with T_{max} values in the PRB
- Look at locations where T_{max} is mapped > 435°C
- Modica and Lapierre (2012) suggest that kerogen hosted porosity adds up to 3% of the bulk rock volume
  - Thermal maturity increases the porosity within the kerogen
- T_{max} calculated at the Field Laboratory correlates well with vitrinite reflectance data collected
  - VR_o of 1.20% and T_{max} of 461°C
  - Late oil window, early wet gas

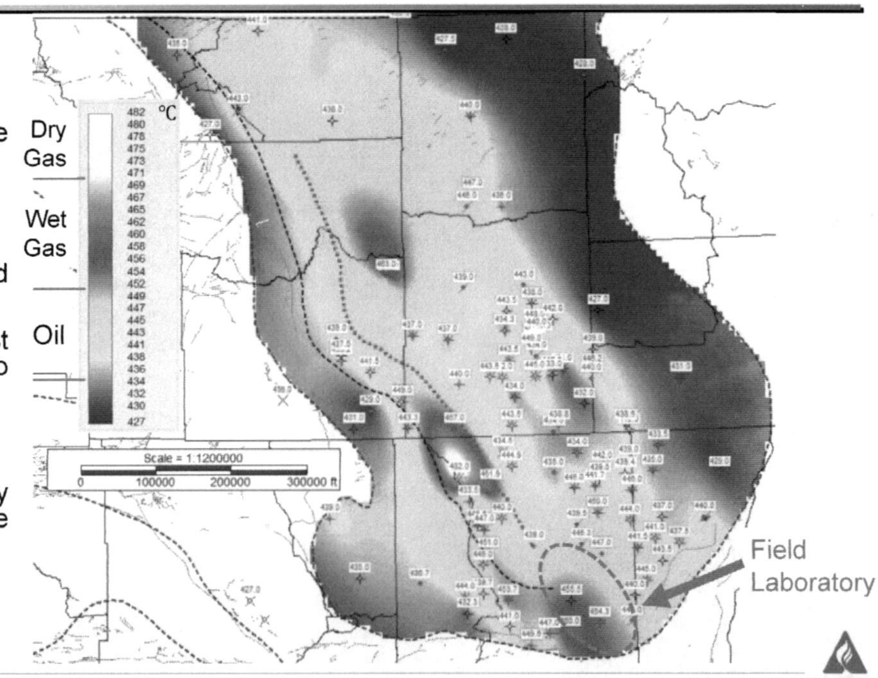

Field Laboratory Mowry Well – Bentonite Flags

- Bentonite layers can act as fluid barriers (not fracture barriers) when hydraulically fracturing the well
  - Try to find lower thickness areas in the PRB for less completions issues
  - Or adjust completions fluid used
  - Issue recognized by University of Wyoming's School of Energy Resources
- Although GR well log resolution is 0.5 feet, the thicker bentonites may cause problems during the frac
  - Layers under 0.5 feet do exist, as seen in core photos/outcrop
- Normalized GR >185 API = bentonite layer
- Confirmed by high Th/K ratio from spectral GR log
  - Thorium concentrations increase in bentonite layers relative to potassium concentrations
- 12.5 feet of total bentonite seen in core photos and confirmed with geophysical logs

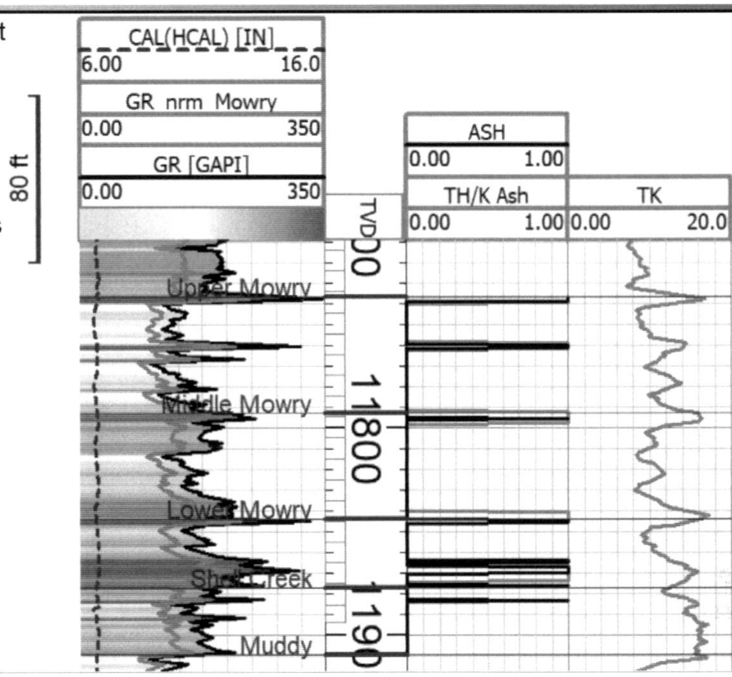

17

Bentonite Layers within the Mowry in the Field Laboratory Well

- Core photos were uploaded into Interactive Petrophysics to compare with normalized GR logs
  - Field Laboratory example shown on the right
    - Two bentonite layers within the Mowry Shale
  - Other PRB Mowry Shale cored wells with photos were uploaded as well to cross check the results throughout the basin
- The Clay Spur Bentonite is a prolific marker bed for the top of the Mowry Shale
  - To quality check the bentonite flag results across the PRB, this marker bed was key

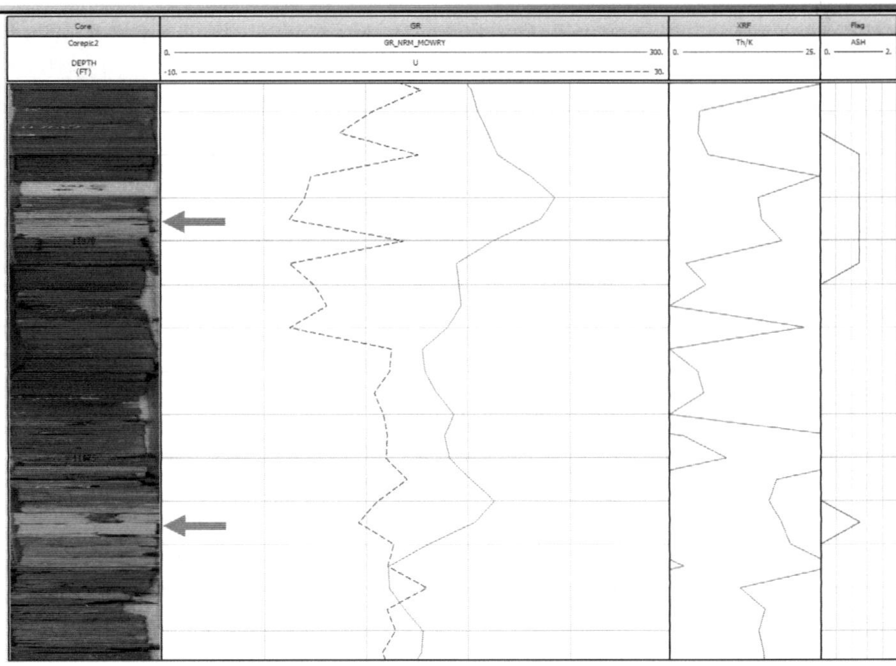

18

Isopach Map of the Bentonite Within the Mowry Shale

- 1,180 wells with normalized GR logs used to generate the map
- General trend of thicker bentonites to the west likely due to volcanic source to the west
- Minimum of ~10 feet of total thickness throughout the Mowry
- Some locations approach 30 feet of total thickness

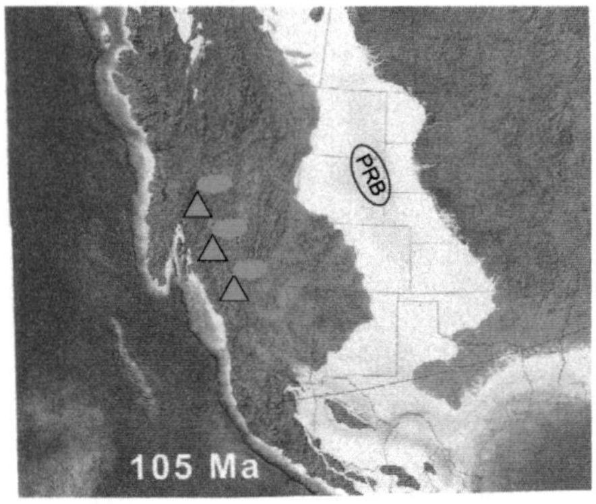

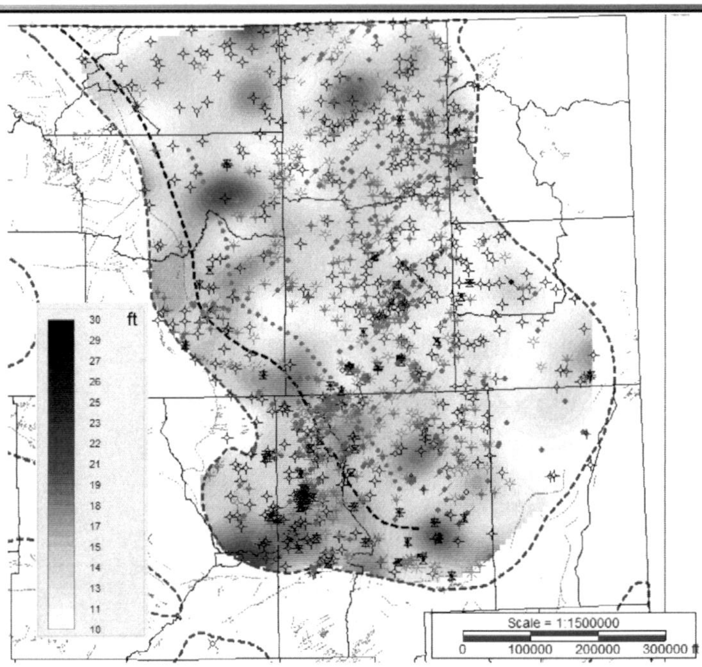

19

Mod. from Blakey and Ranney (2018)

Advanced Resources International, Inc.

Assessing the Mowry Sweet Spots - Thickness (ft) of Middle Member

- 1,180 wells with logs used to create isopach map of the Middle Member
- Trends with Middle Member thickness do not completely align with Total Mowry thickness
  - As seen in Modica and Lapierre (2012)

Total Mowry Thickness (ft)

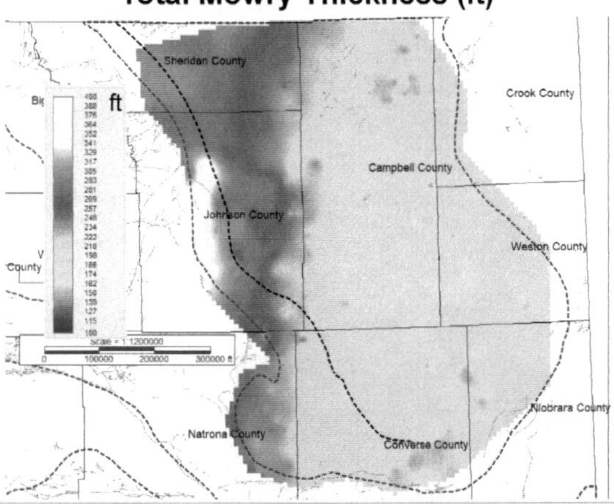

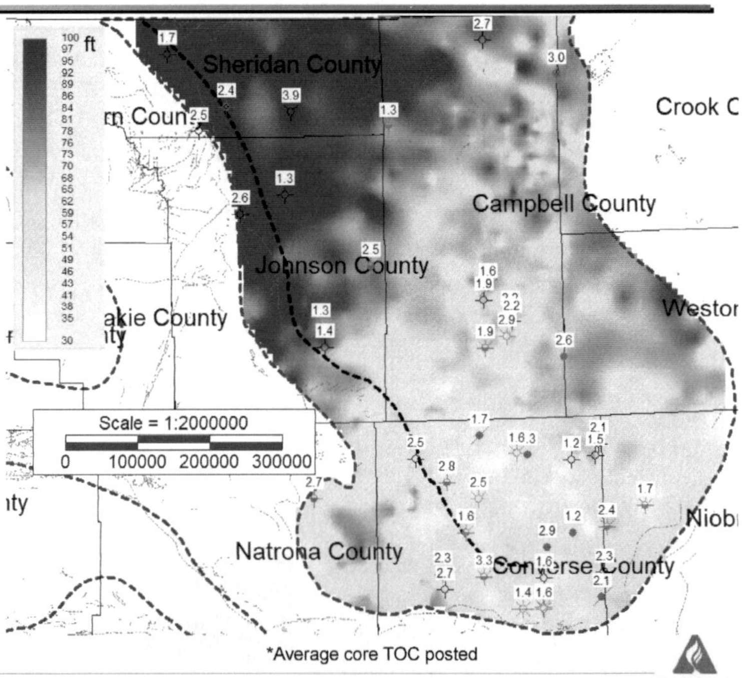

*Average core TOC posted

20

Advanced Resources International, Inc.

Putting it all together - Mowry "Pay"

- Overlay first 6 months of production BOE per 1000 ft lateral
  - All Mowry horizontal wells through 2023
- T_{max} shows the most correlation with production metrics
  - Increased kerogen hosted porosity?

Tmax (°C)

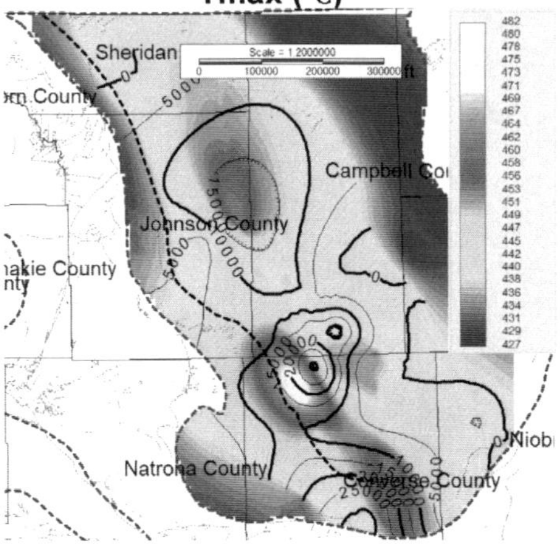

Bentonite Isopach (ft)

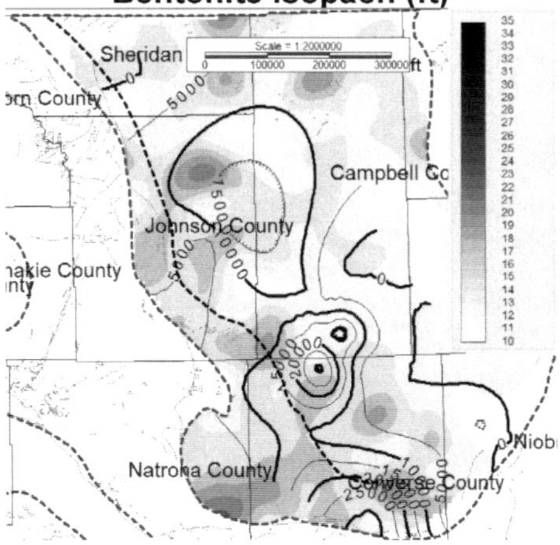

Middle Member Isopach (ft)

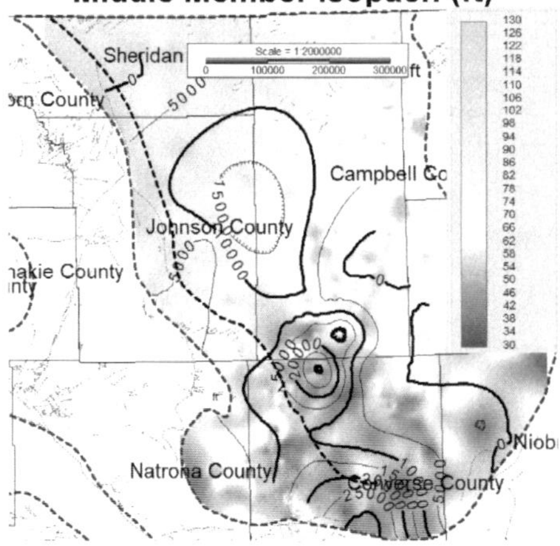

Advanced Resources International, Inc.

Conclusions/Summary

- The Mowry Shale and Turner Sandstone are underexplored, tight-oil reservoirs that still have untapped potential
- The DOE-funded project: *Unlocking the Tight Oil Reservoirs of the Powder River Basin, Wyoming* contributed meaningful data to help in understanding these reservoirs
- The unconventional sweet spots of the Turner Sandstone may exist over thermally mature and overpressured areas in the central PRB
- Adequate evidence to suggest that the Mowry Shale sweet spots exist
  - Thermal maturity and Middle Member thicknesses play roles in understanding sweet spots
  - Need to be aware of bentonite beds for well completions strategy

Turner Pay Zones

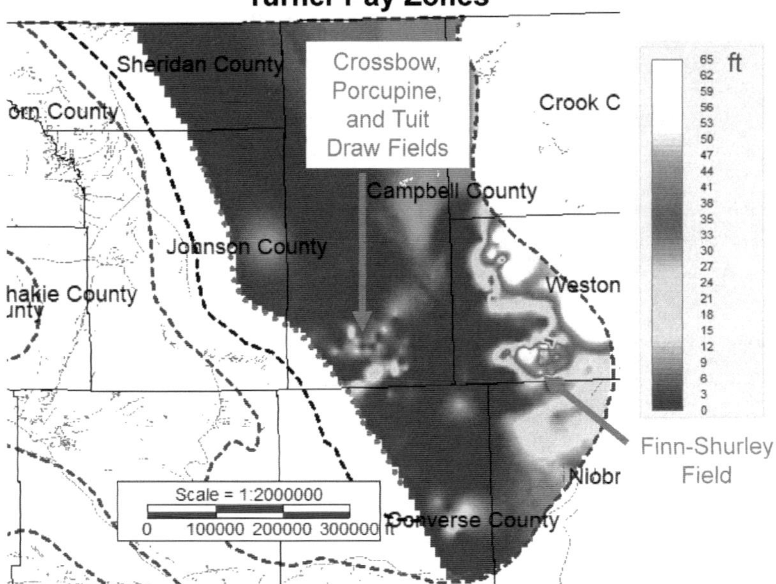

Advanced Resources International, Inc.

Future Work

- Overlay horizontal Turner Sandstone production metrics on pay maps
- Utilize further petrophysical analysis to build out property models
  - Geomechanical properties, absolute and relative permeability
- Create a development strategy for the Mowry Shale, Turner Sandstone, and Niobrara Formation
- Other options for Mowry Shale "pay":
  - RHOB logs: 2.52 g/cm^3 roughly equates to Sw of 60% and TOC of 2%
    - Anything less than or equal to that could be considered "pay"
    - Initial basinal look does not show a correlation like our field laboratory
  - Sonic logs to correlate with geomechanical properties

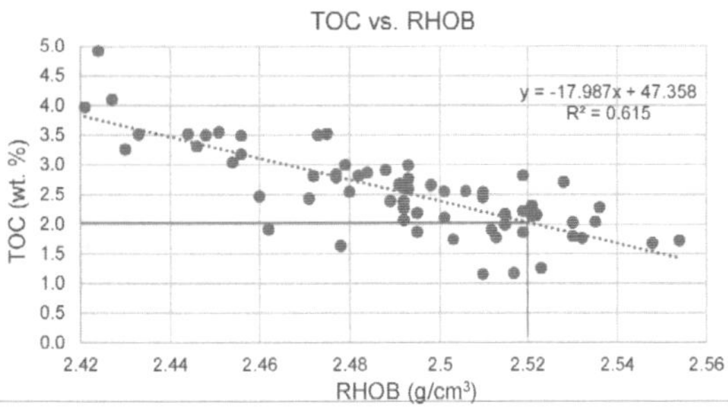

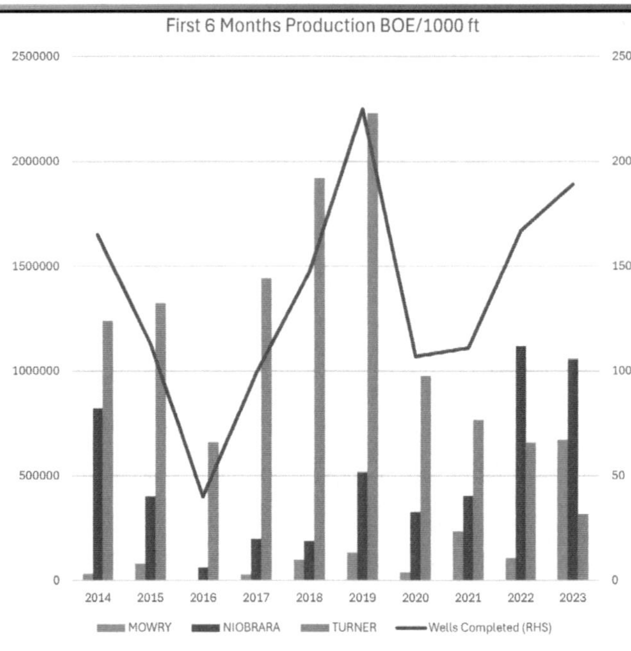

Data from Enverus Prism

23

Advanced Resources International, Inc.

Office Locations

Washington, DC
4501 Fairfax Drive, Suite 910
Arlington, VA 22203
Phone: (703) 528-8420

Knoxville, TN
1210 Kenesaw Ave.
Suite 1210A
Knoxville, TN 37919-7736

Columbus, OH
1840 Mackenzie Dr.
Suite 100
Columbus, OH 43220

| www.adv-res.com

References

- Anna, L., 2009, Geologic assessment of undiscovered oil and gas in the Powder River Basin Province, Wyoming, and Montana: US Geological Survey Digital Data Series, DDS–69–U.
- Blakey, R.C. and Ranney, W.D., 2018, Ancient Landscapes of Western North America: Springer, p. 123.
- Dellenbach, J. T., 2019, Reservoir Characterization and Petroleum Potential of the Upper Cretaceous Wall Creek Member of the Frontier Formation, Western Powder River Basin, Wyoming: unpublished MS thesis, Colorado School of Mines.
- Finley, G. D., 2021, Report on Mowry Shale Thermal Maturity Mapping in the Powder River Basin, Wyoming: EORI Bulletin #2102/2021, *Enhanced Oil Recovery Institute.*
- Heger, A., 2017, Stratigraphy and Reservoir Characterization of the Turner Sandstone, Southern Powder River Basin, Wyoming: unpublished MS thesis, Colorado School of Mines.
- Lichtner, D. T., Toner, R. N., Wrage, J. M., and Lynds, R. M. (2020). Upper Cretaceous Strata in the Powder River Basin: Formation Tops Database, Structure and Thickness Contour Maps, and Associated Well Data, *Open File Report 2020-9, Wyoming State Geological Survey,* 1-55.
- Milar, C. V., 2021, Petrophysical Characterization of Bioturbated Reservoir Facies in the Turner Sandstone, Porcupine and Tuit Draw Fields, Southern Powder River Basin, Wyoming: unpublished MS thesis, Colorado School of Mines.
- Murray, B., Ness, R., Koperna, G., and Carpenter, S. (2021). Evaluation of "Tight Oil" Well Performance and Completion Practices in the Powder River Basin – "Time Slice" Analysis, *URTeC Conference Paper*, 1-18.
- Purvis, S.V., Iwobi, C., Kenny, R.M., Fenton, J.P., Pandey, V., and Davies, C.J., 2017. Facies control on the prospectivity of the unconventional Mowry Formation, southern Powder River Basin, Wyoming, USA. Unconventional Resources Technology Conference (URTeC), DOI 10.15530/urtec-2017-2688030.
- Sonnenberg, S.A., 2020, Carrier Bed (Halo) Play in the Cretaceous Turonian Turner Sandstone, Crossbow Field Area, Powder River Basin: Unconventional Resources Technology Conferences (URTeC), Austin, Texas, USA, 20-22 July 2020.
- Toner, R. N. (2019). Influences on Oil and Gas Production from the Wall Creek and Turner Sandstone Reservoirs, Powder River Basin, Wyoming, *Report of Investigations 77, Wyoming State Geological Survey*, 1-94.

Advanced Resources International, Inc.

Road Traffic Density Analysis and Multi-Objective Optimization of Coal Hauling Trucks Selection by Considering Cost and CO_2 Emissions towards Cleaner Mining Production: Case Study of PT ABC

Irfan Aji Ramadzan[1], Firly Rachmaditya Baskoro[1, 2], Tri Karian[1, 2], Fadhila Achmadi Rosyid[1, 2], Arjo Prawoto Wibowo[1, 2]

[1]Mining Engineering Department, Faculty of Mining and Petroleum Engineering, Institute Technology Bandung, Bandung 40191, Indonesia
[2]Center of Research Excellence of Policy and Economics of Coal and Minerals, Faculty of Mining and Petroleum Engineering, Institute Technology Bandung, Bandung 40191, Indonesia

ABSTRACT

Coal transportation from the coal preparation plant (CPP) to the port is one of the most critical mining activities. One way to transport coal is by truck. PT ABC is a coal company that transports coal from the CPP to the port using several types of trucks. If there is an increase in productivity, the number of trucks will increase. The increased number of trucks will impact the density of coal transportation roads. According to KEPMEN 1827.K/30/MEM/2018, the implementation of transportation must not exceed 80% of the capacity of the transportation route. Therefore, optimizing the type and number of trucks is necessary to minimize coal transportation costs and CO_2 emitted from the transportation activity. The epsilon-constraint method is used to optimize the type and number of trucks in a scenario of 20 – 50 million tons of coal production per annum, with the type and number of trucks determined based on the focus of the decision maker's interests on costs or CO_2 emissions. The optimization results for minimum production of 20 – 50 million tons show that the types of trucks selected are the FH16 truck (120 tons) and the R260 A6 X 4HZ truck (180 tons). Costs and CO_2 emissions have an inverse relationship. When costs are minimal, CO_2 emissions exceed the minimum value, and vice versa.

INTRODUCTION

Coal must go through several stages before reaching consumers. Coal obtained from the mining front will be crushed into a specific desired size at the coal processing facility (coal preparation plant). Coal in the coal preparation plant (CPP) is then transported to the port using several methods, one of which is by using trucks. The plan to increase production by PT ABC has increased the need to haul trucks from the CPP to the port. Based on the Minister of Energy and Mineral Resources Decree Number 1827. K / 30 / MEM / 2018, transportation activities must not exceed 80% (eighty percent) of the capacity of the transportation route. PT ABC, which plans to increase production, will impact the increasing number of trucks so that traffic density on the hauling road will increase. The increased number of trucks will affect transportation costs and CO_2 gas emissions. Optimization uses the epsilon constraint method to select the number of trucks on the transportation road to minimize expenses and CO_2 gas emissions.

LITERATURE STUDY

Road density or traffic density (k, unit/km) is the number of vehicles on a specific length segment of road. Spacing (s, km/unit) is the distance between two cars on the same road lane (Green & Lewis, 2020).The relationship between traffic density and distance can be seen in equation (1).

$$k = \frac{1}{s} \quad (1)$$

Determining spacing can use a safe distance calculation consisting of four components. The first component (D_1) is the distance traveled by the vehicle during the time from the driver's response to being aware and pressing the brake pedal. The first component can be calculated using equation (2).

$$D_1 = V_0 \times DPRT \quad (2)$$

D_1 : First component of braking distance, m
V_0 : Vehicle speed, m/s
$DPRT$: Driver perception and reaction time, s

The second component (D_2) is the distance traveled by the vehicle during the response time of the vehicle's braking system. The final vehicle speed (V_b) is calculated using equation (3). The second component can be calculated using equation (4).

$$V_b = V_0 + (g \times g_{eff} \times BSRT) \quad (3)$$

V_b : Vehicle speed when the braking system responds, m/s
g : gravitational acceleration, m/s^2
g_{eff} : effective grade of the road (grade minus rolling resistance), decreasing as positive, increasing as negative
$BSRT$: Brake system response time, s

$$D_2 = \frac{(V_0+V_b)}{2} \times BSRT \quad (4)$$

The third component (D_3) is the distance required to brake the vehicle to a stop, which can be calculated using equation (5). The fourth component (D_4) is the distance added to consider the safety factor.

$$D_3 = \frac{V_b^2}{2g\times(f-g_{eff})} \quad (5)$$

f : coefficient of friction of the road and road surface

Meanwhile, the CO_2 gas emissions from the use of trucks can be estimated from the fuel consumption of each type of truck. The mathematical equation used can be seen in equation (6) (Kecojevic & Komljenovic, 2010).

$$CO_2 = FC \times CF \quad (6)$$

FC is fuel consumption (L/hour), and CF is a conversion factor of 0.00268 based on the Environmental Protection Agency (Environmental Protection Agency, 2005).

METHOD

PT ABC has a hauling road from the coal preparation plant (CPP) to the port that is 33 km long. The location of the CPP is relatively northwest of the port and is at a higher elevation than the port. The appearance of PT ABC's hauling road can be seen in **Figure 1**. The number and productivity of trucks can be seen in **Table 1**. Total spacing is determined from the safe distance and length of the vehicle shown in equation (7).

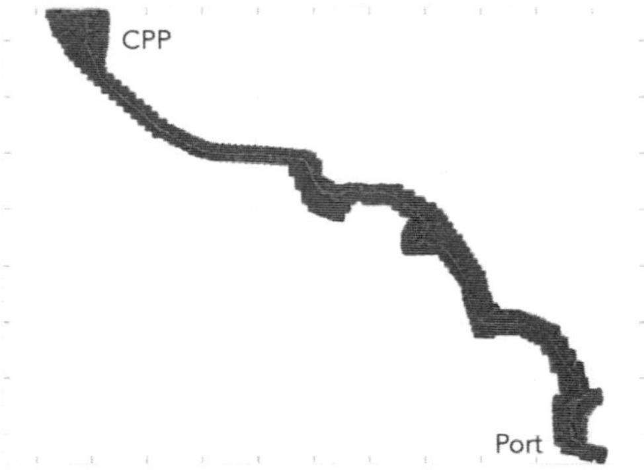

Figure 1. The appearance of PT ABC Hauling Road

Table 1. Truck Number and Productivity

No	Type	Amount	Capacity (Ton)	Productivity per unit (Ton/Year)
1	Quester CGE 370	8	30	50,975.00
2	SYZ440C-8W	22	40	65,262.99
3	P460 LA6X4ESZ	5	75	108,435.07
4	FH16	14	120	162,728.89
5	R580LA-6X4ESZ	52	120	161,210.09
6	R620 A6 x 4HZ STD	14	120	161,927.06
7	R620 A6 x 4HZ	34	180	179,816.55
Total		149		

$$C = \frac{\sum_{i=1}^{n}\sum_{j=1}^{m} x_{ij}(s_{ij_1} + s_{ij_2} + 2d_i)}{L} \times 100\% \quad (7)$$

C is the percentage of road capacity
x_i is the number of trucks for the i-th type
s_{ij_1} is the safe distance for empty conditions on truck-type i and on-road segment j
s_{ij_2} is the safe distance for load conditions on truck type i and on-road segment j
d_i is the length of the i-th type of truck
L is the total length of the round-trip road of 66 km.

Truck operating costs can be calculated by combining fuel, repairs, spare parts, and tire purchase costs. Table 2 shows each type of truck's operating costs, capital, and CO2 emissions.

Table 2. Operating Costs, Truck Capital, and CO2 Emissions per Year

No	Type	Capital Cost (Million Rp)	Operating Cost (Billion Rp)	CO_2 Emission (Ton)
1	Quester CGE 370	650	9.693	724.86
2	SYZ440C-8W	700	10.901	829.91
3	P460 LA6X4ESZ	700	12.723	1,003.25
4	FH16	800	14.941	1,158.20
5	R580LA-6X4ESZ	1.346	15.590	1,121.43
6	R620 A6 x 4HZ STD	1.346	16.439	1,197.59
7	R620 A6 x 4HZ	1.550	17.107	1,213.35

This study modeled several mathematical equation models, which can be seen in equations (8) to (13).

- CO_2 gas emission function
  $$f_1(x) = \sum_{i=1}^{n}\sum_{j=1}^{m}(a_{ij_1} + a_{ij_2})x_{ij} \quad (8)$$
- Function of transportation costs
  $$f_2(x) = \sum_{i=1}^{n}\sum_{j=1}^{m}(b_{ij_1} + b_{ij_2})x_{ij} \quad (9)$$
- Production function
  $$g_1(x) = \sum_{i=1}^{n} c_i x_{ij} \quad (10)$$
- Total road capacity function
  $$g_2(x) = \sum_{i=1}^{n}\sum_{j=1}^{m}(s_{ij_1} + s_{ij_2} + 2d_i)x_{ij} \quad (11)$$
- Road capacity function per segment with a length of 1 km

$$g_3(x) = \sum_{i=1}^{n}(s_{ij_1} + s_{ij_2} + 2d_i)x_{ij} \quad (12)$$

- The minimum number of trucks in each type is 0.

$$x_{ij} \geq 0 \quad (13)$$

With:

x_{ij} is the number of trucks for the i-th truck type on the j-th road segment

a_{ij_1} is the CO_2 gas emission from the i-th truck type on the j-th road segment in empty condition.

a_{ij_2} is the CO_2 gas emission from the i-th truck type on the j-th road segment in a loaded condition.

b_{ij_1} is the transportation cost of the i-th truck type on the j-th road segment in empty condition.

b_{ij_2} is the transportation cost of the i-th truck type on the j-th road segment in loaded condition.

c_i is the productivity of the i-th truck type.

s_{ij_1} is the spacing or safe distance for the i-th truck type on the j-th road segment in empty condition.

s_{ij_2} is the spacing or safe distance for the i-th truck type on the j-th road segment under load.

d_i is the vehicle length for the i-th truck type.

The scenario starts with a productivity limit of 20 million tons per year, and optimization is carried out using the epsilon constraint method. Then, optimization is carried out for the following scenario by adding a productivity limit of 10 million tons per year, which is more than the previous scenario. Each type's minimum number of trucks is zero, and the road capacity is 80% of the maximum road capacity. The following is a mathematical equation for optimizing using the epsilon constraint method.

Objective function:

- Minimize CO_2 emissions

$$f_1(x) = \sum_{i=1}^{n}\sum_{j=1}^{m}(a_{ij_1} + a_{ij_2})x_{ij} \quad (14)$$

- Minimize transportation costs

$$f_2(x) = \sum_{i=1}^{n}\sum_{j=1}^{m}(b_{ij_1} + b_{ij_2})x_{ij} \quad (15)$$

With the following constraints:

- Trucks are filled to a predetermined capacity.
- The minimum number of trucks in each type is 0

$$x_i \geq 0 \quad (16)$$

- Production target according to scenario

$$g_1(x) \geq \text{Production target} \quad (17)$$

- The number of tools permitted is 80% of the road capacity, with a length of 33 km and 66 km for return.

$$g_2(x) \leq 66000 \times 0.8 \quad (18)$$

- The number of tools permitted per segment with a length of 1 km is 80% of the road capacity per segment.

$$g_3(x) \leq 1000 \times 0.8 \quad (19)$$

The epsilon constraint method limits the objective function to minimize costs with a value. This value is obtained using epsilon, which has a value between 0 - 1 from the difference between the value of transportation costs when CO_2 gas emissions are at a minimum value with the minimum value of transportation costs plus the minimum value of transportation costs. The epsilon value increases incrementally by 0.1 from 0 to 1. The mathematical model for the transportation cost function becomes an equation (20).

$$f_2(x) \leq \varepsilon\left(f_{2\,f_1\min} - f_{2\,\min}\right) + f_{2\,\min} \quad (20)$$

RESULT

The calculation result of the percentage of actual hauling road capacity with a production target of 18 million tons per year produces a value of 40.01%. The optimization results using the epsilon constraint method consist of four scenarios, namely scenario 1 with a minimum production limit of 20 million tons, scenario 2 with a minimum production limit of 30 million tons, scenario 3 with a minimum production limit of 40 million tons, and scenario 4 with a minimum production limit of 50 million tons. The scenario with a minimum production limit of 60 million tons does not produce a solution set, so the scenario stops at scenario 4. The results of the solution set for scenarios 1 - 4 can be seen in **Figure 2**.

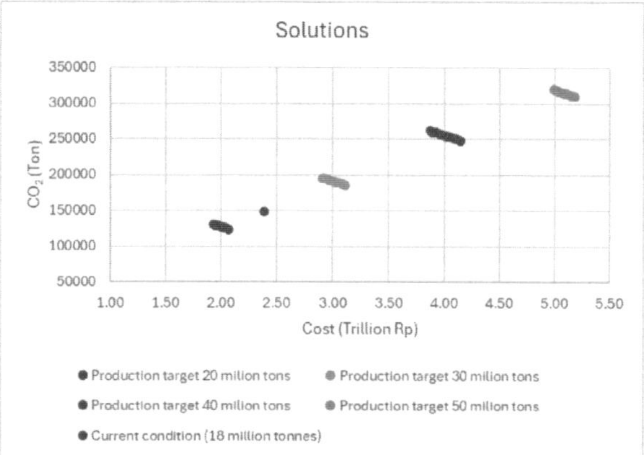

Figure 2. Objective Function Solution Set for Scenarios 1 – 4

In scenario one, with a minimum production target of 20 million tons, several scenarios can be selected, as shown in **Figure 3, Figure 4,** and **Figure 5**.

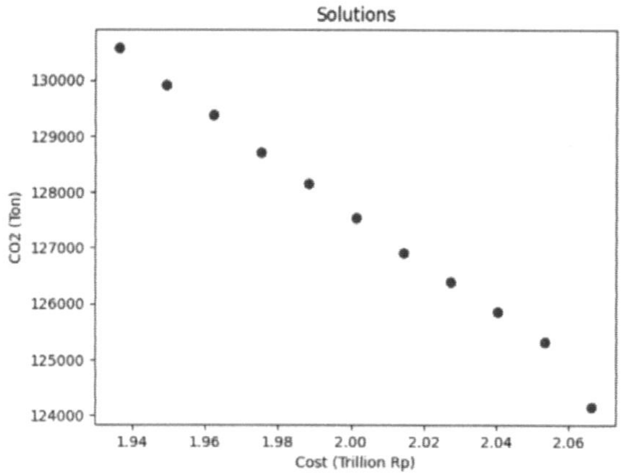

Figure 3. Objective Function Solution Set for Scenario 1

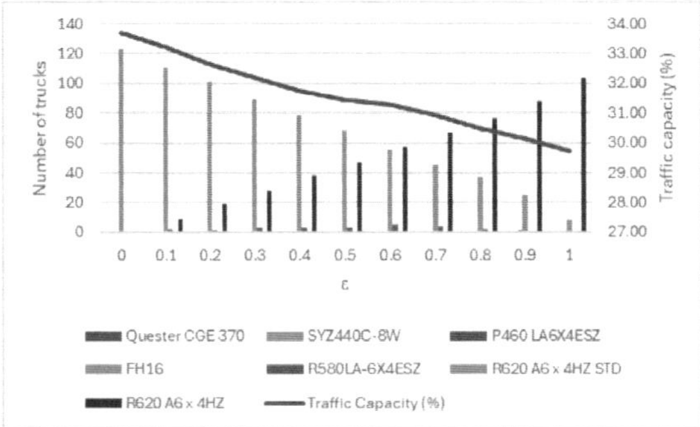

Figure 4. Number of Trucks and Traffic Capacity Scenario 1

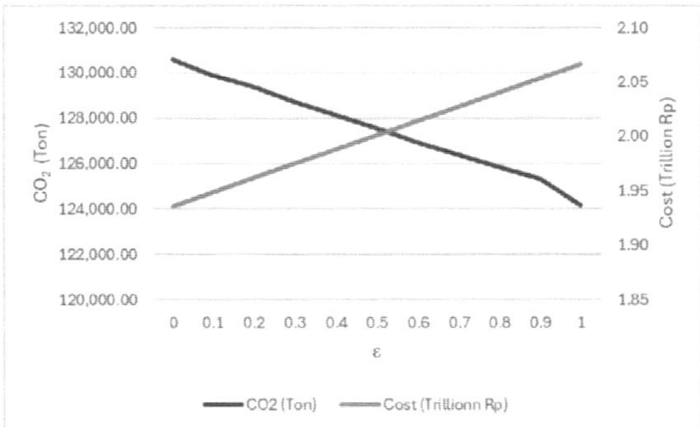

Figure 5. Costs and CO_2 Emissions Scenario 1

In scenario two, with a minimum production target of 30 million tons, several scenarios can be selected, as shown in **Figure 6, Figure 7,** and **Figure 8.**

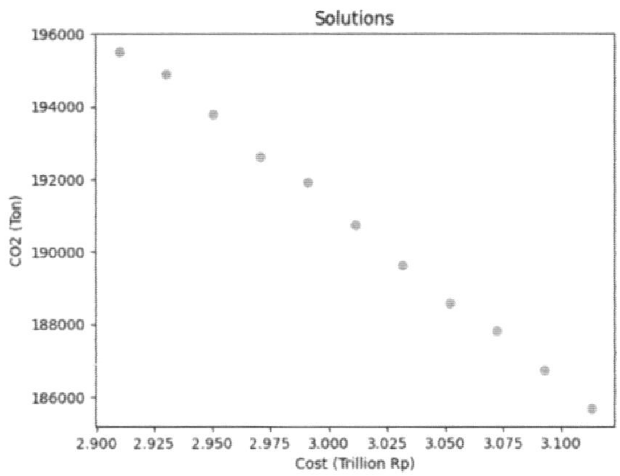

Figure 6. Objective Function Solution Set for Scenario 2

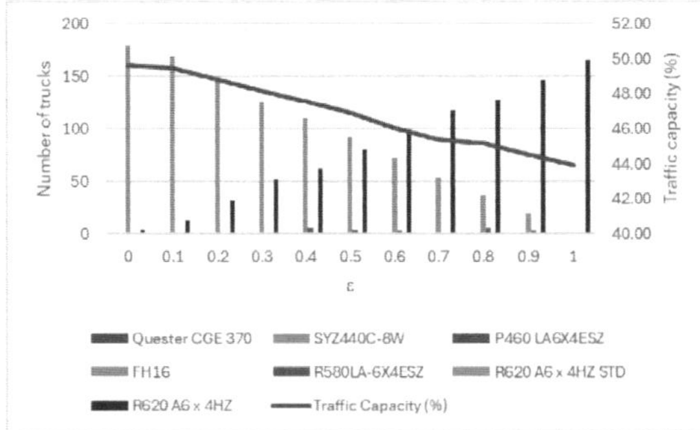

Figure 7. Number of Trucks and Traffic Capacity Scenario 2

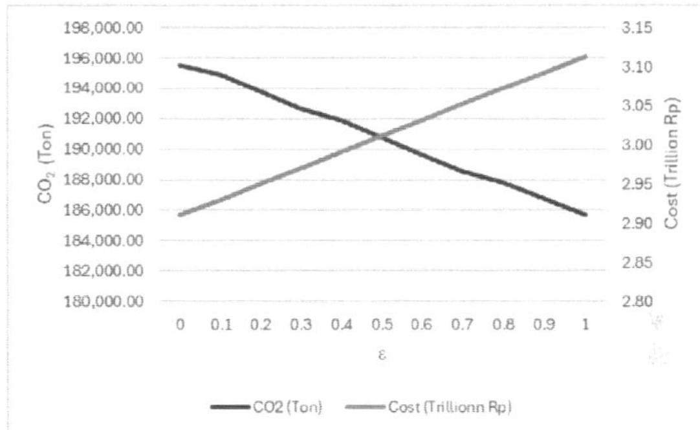

Figure 8. Costs and CO_2 Emissions Scenario 2

In scenario three, with a minimum production target of 40 million tons, several scenarios can be selected, as shown in **Figure 9, Figure 10,** and **Figure 11.**

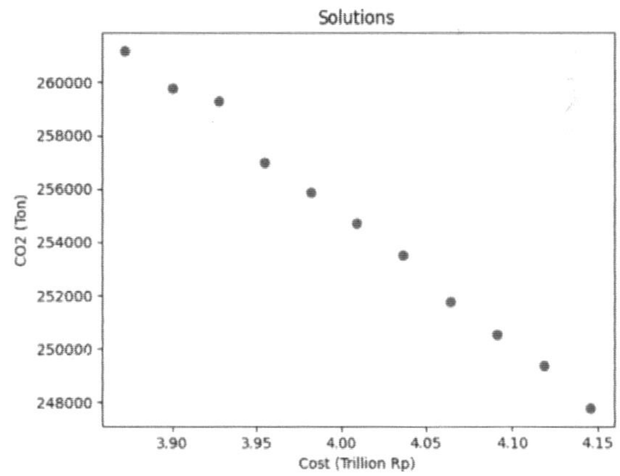

Figure 9. Objective Function Solution Set for Scenario 3

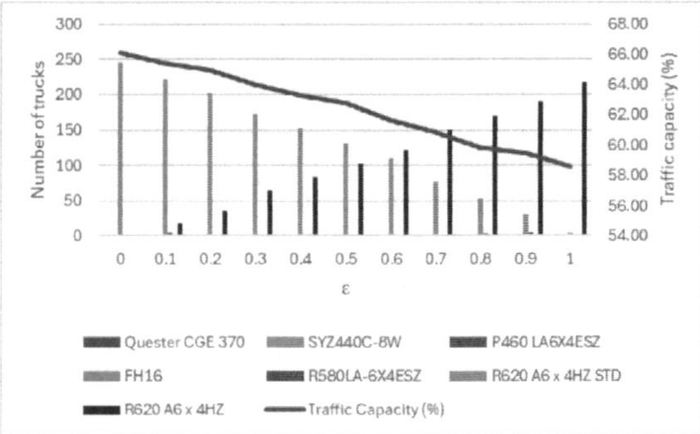

Figure 10. Number of Trucks and Traffic Capacity Scenario 3

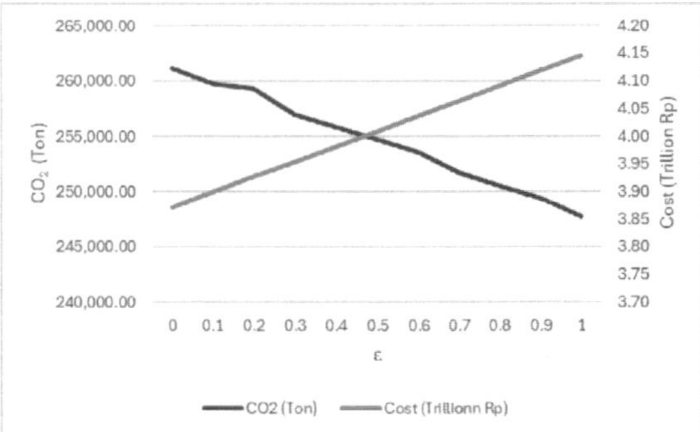

Figure 11. Costs and CO_2 Emissions Scenario 3

In scenario four, with a minimum production target of 50 million tons, several scenarios can be selected, as shown in **Figure 12, Figure 13,** and **Figure 14.**

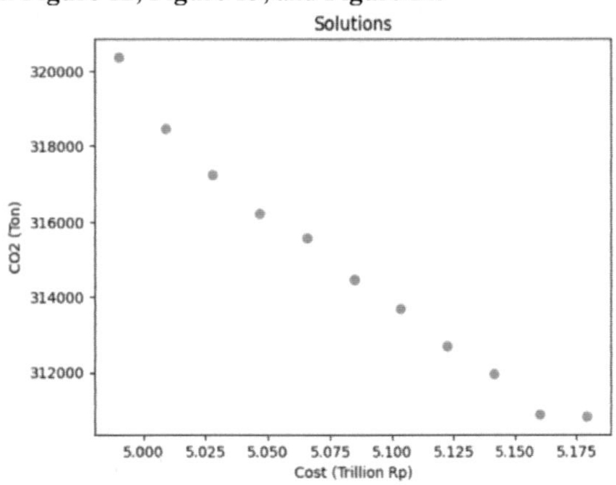

Figure 12. Objective Function Solution Set for Scenario 4

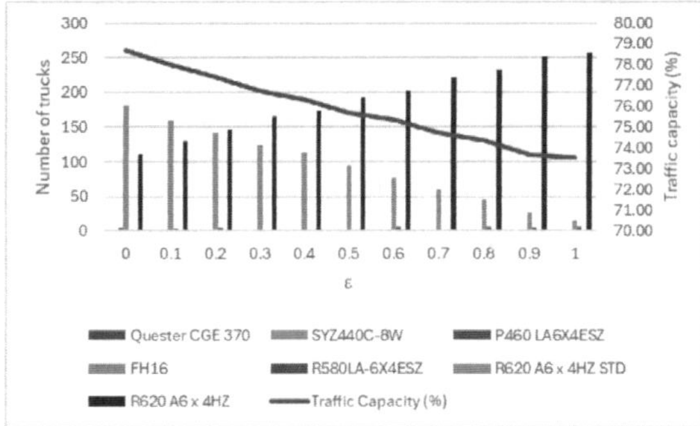

Figure 13. Number of Trucks and Traffic Capacity Scenario 4

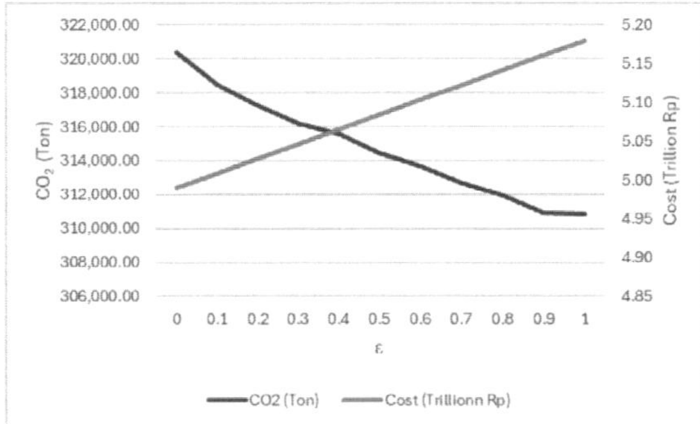

Figure 14. Costs and CO_2 Emissions Scenario 4

The most significant total CO_2 emissions occur if the selected transportation mode is a gasoline truck. Although it produces high CO_2 emissions, the transportation cost is the smallest compared to other modes. Meanwhile, using EV trucks will provide the advantage that there are no CO_2 emissions produced by the trucks themselves. However, the high capital costs of purchasing EV trucks and building facilities such as electric charging and maintenance will result in a total price that will be higher

Table 3. Comparison of Cost and CO_2 Emissions between Gasoline Truck, Belt Conveyor, and EV Truck

Parameter	Gasoline Truck	Belt Conveyor	EV Truck
Total Cost (Billion Rp)	4,950-5,180	6,952.76	8,267.57
CO_2 (Ton)	311,000-320,500	249,058.14	-

DISCUSS

The optimization results for a minimum production of 20

million tons to 50 million tons show that, generally, the types of trucks chosen are the FH16 truck (120 tons) and the R260 A6 X 4HZ truck (180 tons). In 20 million to 40 million tons production conditions, the FH16 truck type (120 tons) is widely chosen when the cost component is more critical to minimize than CO_2 emissions. Meanwhile, the R260 A6 X 4HZ truck type (180 tons) is more widely chosen when the CO_2 emission component is more critical to minimize than cost. The FH16 truck type costs less than the R260 A6 X 4HZ truck type. Meanwhile, the larger capacity of the R260 A6 X 4HZ truck type means fewer trucks are needed, resulting in fewer CO_2 emissions.

The optimization results for a minimum production of 50 million tons generally prefer the R260 A6 X 4HZ type truck. Although the R260 A6 X 4HZ truck costs the most, the number of trucks needed for large production is less. This is also in line with the CO2 emissions produced due to the fewer trucks required compared to using trucks with a smaller capacity.

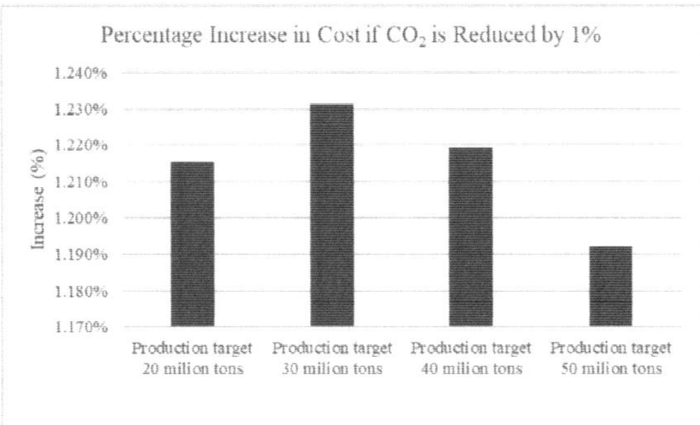

Figure 15. Percentage Increase in Cost if CO_2 is Reduced by 1%

Reducing CO_2 emissions has an impact on increasing transportation costs. In scenario one, with a production target of 20 million tons per year, when reducing CO_2 emissions by one percent, there will be a cost increase of 1.215%. Meanwhile, for scenarios two, three, and four, each will experience a rise of 1.231%, 1.219%, and 1.192% if CO_2 emissions are reduced by one percent, as in **Figure 15.** Reducing CO_2 emissions will result in a higher cost increase at a tiny production target than at a large one.

At a production target of 50 million tons per year, the transportation method with gasoline trucks produces the smallest transportation costs but the most significant CO_2 gas emissions compared to belt conveyors and EV trucks. While the transportation method with EV trucks does not produce CO_2 gas emissions during coal transportation, it should be considered that the transportation costs are the highest.

ACKNOWLEDGEMENTS

Gratitude is expressed to the supervising lecturers, Mr. Firly Rachmaditya Baskoro, Mr. Fadhila Achmadi Rosyid, Mr. Arjo Prawoto Wibowo, as well as parents and friends who have provided support during this research.

REFERENCES

Brandimarte, P. (2014). *Handbook in Monte Carlo Simulation.* Chicago: Wiley.

Elam, R. A., Teaster, E. C., & Lawless, M. J. (1999). *MSHA Handbook Series.* U. S. Department of Labor.

Environmental Protection Agency. (2005). *Average carbon dioxide emission resulting from gasoline and diesel fuel.* Washington DC: Office of Transportation and Air Quality.

Green, D., & Lewis, K. (2020). *Guide to Traffic Management Part 2: Traffic Theory Concepts.* Sydney: Austroads Ltd.

Hillier, F. S., & Lieberman, G. J. (2020). *Introduction to Operations Research.* McGraw-Hill.

Kecojevic, V., & Komljenovic, D. (2010). Haul Truck Fuel Consumption and CO2 Emission Under Various Engine Load Conditions. *Mining Engineering*, 44-48.

Kim, N. S., Janic, M., & Wee, B. v. (2009). Trade-Off Between Carbon Dioxide Emissions and Logistics Costs Based on Multiobjective Optimization. *Transportation Research Record: Journal of the Transportation Research Board*, 107-116.

Mavrotas, G. (2009). Effective implementation of the ε-constraint method in Multi-Objective Mathematical Programming problems. *Applied Mathematics and Computation*, 455-465.

Pusat Data dan Teknologi Informasi Kementrian Energi Dan Sumber Daya Mineral. (2023). *Handbook of Energy & Economic Statistics of Indonesia.* Jakarta: Kementrian Energi Dan Sumber Daya Mineral.

The Australasian Institute of Mining and Metallurgy. (2012). *The Australasian Institute of Mining and Metallurgy: Second Edition, Monograph 27.* Carlton Victoria: The Australasian Institute of Mining and Metallurgy.

Winston, W. L. (1994). *Operation Research: Apllications and Algorithms.* California: Duxbury Press.

2024 International Pittsburgh Energy and Carbon Management Conference

October 7 - 9, 2024

Manuscript Submission

PROGRAM TOPIC:
COMBUSTION TECHNOLOGIES

THERMOGRAVIMETRIC ANALYSIS COMBUSTION CHARACTERISTICS OF COAL MICROLITHOTYPE PARTICLE TYPES

Presenting Author
Lesigen Moodley, Mr, Eskom Research, Testing and Development, Lower Germiston Road, Rosherville, Johannesburg, South Africa,
moodll@eskom.co.za

Co-Authors
Frikkie Conradie, Dr, North-West University, School of Chemical and Mineral Engineering, Potchefstroom, South Africa, frikkie.conradie@nwu.ac.za

Walter Schmitz, Prof, University of Witwatersrand, School of Mechanical, Industrial and Aeronautical Engineering, Johannesburg, South Africa, walter.schmitz@wits.ac.za

Hein Neomagus, Prof, North-West University, School of Chemical and Mineral Engineering, Potchefstroom, South Africa, hein.neomagus@nwu.ac.za

Bonny Nyangwa, Mr, Eskom Research, Testing and Development, Lower Germiston Road, Rosherville, Johannesburg, South Africa, nyangwtb@eskom.co.za

Olifant Sinthumule, Mr, Eskom Research, Testing and Development, Lower Germiston Road, Rosherville, Johannesburg, South Africa, sinthuom@eskom.co.za

Abstract:

This study examined the impact of coal microlithotype particle type proportion on combustion characteristics using a Thermogravimetric Analyser (TGA). Coal samples evaluated were enriched in various microlithotypes including Vitrite80, Vitrite60, VITRSF, VITSF and FUS. The TGA combustion results were post-processed to determine the sample ignition, burnout, peak temperatures and reactivity. Findings revealed that ignition, burnout, peak temperatures and reactivity during combustion is influenced by the concentration of certain coal microlithotype particle types. This study highlights the significance of the proportion of various microlithotype particle types in coal to achieve lower ignition and burnout temperatures and enhanced reactivity which would influence boiler performance.

Keywords: coal microlithotype particle types, Scanning Electron Microscopy, Thermogravimetric Analysis, combustion characteristics, Float-Sink processing

1. Introduction

Coal is an important energy source for power generation [1, 2, 3, 4]. However, increasing environmental concerns surround coal mining operations and the associated emissions released during coal combustion [5, 6, 7]. Additionally, declining coal quality at power stations [1, 8] which has contributed to boiler combustion related issues such as primary ignition problems [9] slagging [10, 11], fireside corrosion [8], temperature excursions [10] and high unburnt carbon losses [12]. To ensure coal's continued use in power generation, it is crucial to minimise its environmental impact and address these operational challenges. This requires a comprehensive understanding of the combustion characteristics of new coal sources before they are fired in industrial boilers.

Computational Fluid Dynamics (CFD) simulations are widely used to assess combustion characteristics of new coal sources and blends [13], optimise the efficiency of energy systems [14], simulate changes to operating parameters [15, 16] and burner modifications [17]. Additionally, CFD models can track coal particle interactions [18] and simulate flow patterns within boiler and model key combustion characteristics such as coal ignition, burnout, particle residence time and also simulate gaseous species formation e.g. Oxides of Nitrogen (NO_x) [19] and Sulphur Oxides (SO_x) and provide insights into temperature distribution and hot spots to minimize slagging formation. By integrating chemical kinetics, CFD allows for the simulation of complex chemical reactions during combustion, enabling the testing of different scenarios without the need for costly physical experiments. This allows for the identification of potential problems early and the testing of potential solutions prior to firing and implementing in industrial boilers. However, while CFD has demonstrated accuracy in predicting the flow field [20] and gas temperatures during coal combustion, other predicted quantities such as gaseous emissions and burnout levels have shown to be less accurate [21]. Burnout predictions are particularly important since they are utilised to estimate the boiler furnace height during the design of new units and to evaluate potential use of new coal sources for existing boilers. These predictions inaccuracies may stem from the complex nature of coal and the inherent challenges in modelling its combustion behaviour [22, 23, 24]. Therefore, although CFD remains a valuable tool in coal combustion studies, further refinement of certain model outputs particularly emissions and burnout are necessary to improve prediction reliability.

Coal combustion behaviour is influenced by both external and inherent factors, which need to be included in Computational Fluid Dynamics (CFD) models for reliable predictions [25]. External factors, such as particle size, air temperature, burner swirl setting, design and furnace height, have been studied extensively [26]. In contrast, inherent factors, including the maceral and mineral content of coal, its rank, porosity, reactivity, exposed surface area, particle size, and degree of weathering, are equally significant but more challenging to model [25]. The heterogeneous nature of coal, especially after pulverization, introduces substantial variability in particle composition [27], which must be accounted for to ensure accurate combustion models. Macerals, the organic constituents of coal, influence physical and chemical properties, including reactivity and burnout behaviour, while mineral matter contributes to slagging. Pulverized coal samples are particularly heterogeneous, containing particles with vastly different maceral compositions, which behave differently upon heating. For instance, variations in maceral constituents can affect the yield of volatile matter, char reactivity, and ash chemistry [28, 29, 30], leading to different combustion outcomes for coal particles within the same sample.

Gibbins et al. [31] noted that conventional coal analyses are not well-suited to describe this heterogeneity. Traditional methods assume coal is homogeneous and are focused on obtaining average values, which may not reflect the true behaviour of a heterogeneous particle population. An alternative method based on maceral associations should be used to describe heterogeneity in pulverised coal particles. Average coal properties are appropriate for models such as overall plant Mass and Energy Balances (MEB). Additionally, the combustion behaviour of a heterogeneous array of coal particles can significantly differ from that of a homogeneous sample with the same average composition. In particular, individual particle properties are crucial in gas-solid char combustion, influencing the degree of unburnt carbon leaving the boiler. Therefore, understanding and incorporating coal heterogeneity into combustion models, particularly CFD simulations, is essential for accurately predicting ignition, burnout, and emissions under industrial conditions.

Eskom has undertaken extensive research to understand the heterogeneity of pulverized coal particles and their impact on power station performance, particularly in relation to slag formation, high unburnt carbon, flame stability, and elevated furnace exit gas temperatures. This research led to the development of a pioneering approach by Eskom's Research, Testing, and Development (RT&D) department, utilizing **Q**uantitative **E**valuation of **M**inerals by **Scan**ning Electron Microscopy (QEMSCAN) to analyze coal particles [32]. QEMSCAN produces detailed false-color particle maps, distinguishing between homogeneous particles, such as those composed solely of macerals e.g. Vitrinite or minerals e.g. Quartz, Kaolinite, and heterogeneous particles containing various maceral and mineral associations in differing proportions. Furthermore, to build on the Scanning Electron Microscopy (SEM) analysis, RT&D introduced the Microlithotype Particle Classification System (MPCS), which comprises 49 distinct particle types [32, 27]. This classification expands on traditional classification used in petrographic analysis [33] by grouping coal particles into three categories: maceralic, carbominerites, and minerites. The maceralic group includes particles rich in macerals, such as Vitrite80, containing ≥80% (by mass) Vitrinite, or mixed particles like VITRSF, which contains ≥60% Vitrinite and Reactive Semifusinite. The carbominerites group encompasses coal-mineral combinations, such as carbosilicate_MM40, which contains coal and quartz ≥40%. Finally, the minerites group is composed primarily of minerals, like Pyrite Cleat, which contains ≥60% Pyrite. This advanced particle classification system allows RT&D to quantify the distribution of microlithotype particle masses in pulverized coal samples from South Africa and internationally, revealing significant differences in their mass distributions [32, 27]. In addition, RT&D has adapted a **TES**la **SCAN**ning (TESCAN) Scanning Electron Microscope to quantify microlithotype particle distribution [34]. While this work has significantly advanced the characterization of coal particle heterogeneity, further research is needed to explore the combustion characteristics and interactions of these coal microlithotype particle types, which remain poorly understood.

Previous CFD studies applied a maceral correction factor to improve the accuracy of char burnout models by adjusting for the influence of maceral content on reactivity [35, 14]. However, these studies primarily focused on average properties, simulating coal particles with similar maceral compositions and, consequently, similar char reactivity. While the significance of coal inter-particle heterogeneity is well recognized, its effects have not yet been integrated into CFD combustion models. This omission limits the ability of current models to make reliable predictions regarding combustion behaviour. Capturing this heterogeneity is essential for improving the accuracy of combustion predictions, particularly in relation to ignition, burnout, and slagging phenomena. An understanding of the heterogeneous nature of pulverized coals is fundamental to the development of more precise conceptual and quantitative models for coal processing and combustion [31]. By incorporating inter-particle heterogeneity into CFD models, researchers can better simulate the complex interactions between maceral and mineral constituents, ultimately leading to improved plant optimization, enhanced blending strategies, and more effective environmental mitigation efforts. Therefore, further research is necessary to determine the combustion characteristics of coal inter-particle maceral and mineral heterogeneity and to integrate these features into CFD models for more accurate and reliable combustion simulations.

This study determined the Thermogravimetric Analysis (TGA) combustion characteristics, specifically ignition, burnout and peak temperatures, as well as the reactivity of coal microlithotype particle-type concentrates.

2. Materials and Methods

2.1. Sample origin and Float-Sink processing

A South African export coal sample, vitrinite-rich and low-ash yield from the Witbank Coalfield was selected for this study. This sample was chosen because its lower ash yield as compared to local power station coals, makes it more efficient to concentrate microlithotype particle types. The bulk coal sample was crushed to a size of -1 mm. Float-sink separation of the coal sample using tight density ranges was conducted at Bureau Veritas Laboratories in Centurion.

The dense liquid for separation (1.30–1.62 g/cm³) was prepared by blending perchloroethylene (1.62 g/cm³) with white spirits (0.79 g/cm³). Liquid density was verified using a pycnometer or a calibrated measuring cylinder and a calibrated scale. Heavy liquids such as zinc chloride were avoided to prevent pore adhesion, which could interfere with subsequent coal analysis [36, 37]. The selected liquid densities ranged from 1.30 g/cm³ for pure vitrinite to 1.50 g/cm³ for inertinite [38].

A 500 g coal sample was mixed with the dense liquid in a 25-liter container and manually agitated for 45 minutes. Next, the mixture was poured into a separation funnel and allowed to settle for up to 24 hours, the float fraction was collected using filter paper, rinsed, and air-dried. The sink fraction underwent further processing to yield the final fractions [39]. The Float-Sink processing produced 12 sample fractions. Additional procedural details on the Float-Sink separation method are outlined in Moodley et al. [40].

2.2. SEM analysis and coal microlithotype particle type classification

SEM sample preparation involved subdividing each coal samples into ~0.5 g portions using Quantachrome's micro riffle. Each portion was mixed with molten carnauba wax and moulded into 30 mm diameter blocks [41, 27]. Carnauba wax was chosen because it allows for the analysis of both organic and inorganic constituents, thereby facilitating clear differentiation between coal particles from the mounting medium. The samples were polishing and carbon coated to reduce surface roughness and charge issues [42]. SEM analysis was conducted using Eskom RT&D's TESCAN TIMA instrument. During the analysis, Backscattered Electrons (BSE) and Energy Dispersive X-Rays (EDS) produced were compared with Species Identification Protocol (SIP) file for phase identification. Once all points were analysed on individual coal particles, the system generated false-colour particle maps which illustrated the maceral and mineral distribution [32]. Verification SEM results involved a comparison of total mineral matter content with proximate ash yield.

TIMA software was utilised to classify coal particles in accordance with Eskom RT&D's Microlithotype Particle Classification System (MPCS), following the novel methodology developed by van Alphen [34]. Verification of SEM results involved comparison with proximate ash yield results. The SEM data, including particle maps and microlithotype particle type distribution data, were exported to Excel format for subsequent processing. This allowed for the annotation of particle maps and the creation of graphical representations illustrating the distribution of microlithotype particle types.

2.3. TGA combustion characterization of coal microlithotype particle types

TGA sample preparation involved pulverizing the raw coal and derived microlithotype concentrates to -212 μm. Each sample was subjected to TGA combustion analysis using Eskom RT&D's LECO 701 instrument, following the process outlined in **Fig. 1**. Approximately 100 mg of each sample was loaded into a ceramic crucible for the analysis. Based on the recommendations provided by Bada [43], the TGA tests were conducted at a constant heating rate of 5 °C/min, with a final temperature of 900 °C. Synthetic air comprising (21% Oxygen and 79% Nitrogen by volume) was supplied at a consistent flow rate of 3.5 L/min. The combustion program was completed once mass consistency was achieved. During the experiments, the TGA system continuously recorded the sample's weight loss, furnace temperature and time. TGA output data was exported to Excel format and included 1st and 2nd derivatives of weight loss versus time. Peak temperature and maximum reactivity were determined using the method described by Bada et al. [44]. The intersection method, as outlined by Chao, et al. [45], Lu & Chen, [46] and Isaac [47] was used to determine the sample's ignition temperature while burnout temperature was established at ~99% conversion, following the procedure described by Lu & Chen [46].

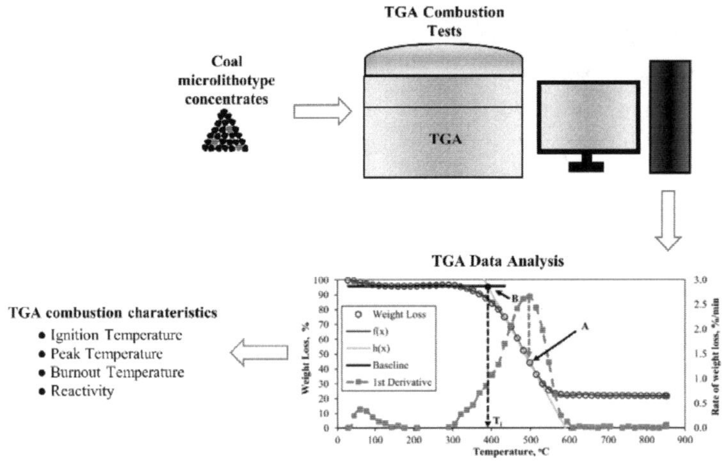

Fig. 1. Coal microlithotype particle type TGA combustion test process

2.4. Correlation of TGA combustion characteristics with coal microlithotype particle type mass distribution

TGA combustion characteristics results were correlated with coal microlithotype particle type mass distribution to assess any potential relationships. This process was automated using a Python script. Logarithmic feature transformation of the microlithotype mass concentrations was applied to account for non-linear patterns and improve the interpretability of the data. The natural logarithm of the microlithotype concentrations was plotted against the TGA combustion characteristics, and potential outliers were identified and removed using the interquartile range (IQR) method. Linear regression was subsequently performed, with the goodness-of-fit evaluated by calculating the R-square value.

3. Results & Discussions

3.1. Scanning Electron Microscopy results

SEM results were validated by comparing the total mineral matter content with proximate analysis ash yield, revealing strong agreement and confirming the SEM reliability. SEM particle maps for raw coal and the derived float-sink fractions are shown in **Fig. 2 (a) to (m)**. The raw coal exhibited particles with varying maceral and mineral proportions. Various particle types were identified, each with distinct characteristics. Some were mainly Vitrinite-rich, associated with Fusinite and Kaolinite (Particles 1, 8, 11), while others had Vitrinite with Reactive Semifusinite (RSF) inclusions (Particles 2, 6, 9). Some particles were composed mostly of Kaolinite with minimal Vitrinite (Particles 3, 4, 7), while others contained Vitrinite, Semifusinite, and Kaolinite (Particle 5). Particles with Vitrinite and Dolomite (Particle 10), Vitrinite mixed with Fusinite, Semifusinite, RSF (Particles 12, 21), and Fusinite with Vitrinite, Semifusinite, Kaolinite (Particle 13) were also observed. Mineral-rich particles contained Pyrite (Particle 16). These maps provide insights into complex maceral-mineral associations in the raw coal particles.

The obtained float-sink fractions were S1.275-F1.30 to S1.600. Particle maps from these fractions revealed insights into their composition **Fig. 2 (b) to (m)**. S1.275-F1.30 was mainly Vitrinite with RSF inclusions (Particle 1). S1.30-F1.325 had Vitrinite with RSF, Fusinite, and Kaolinite (Particle 1). S1.325-F1.35 showed vitrinite-rich particles (Particles 12, 16, and 18) and Vitrinite particles with Fusinite, Quartz, Kaolinite (Particles 1, 2). S1.35-F1.425 contained Vitrinite with varied associations, including Dolomite, Kaolinite and Semifusinite (Particles 3, 4, 5, 6, 8, 9). Mineral content increased in this fraction. S1.425-F1.45 had Vitrinite with Fusinite, Kaolinite, and Dolomite (Particles 1, 2). Fractions S1.457-F1.575 to S1.600 had higher Kaolinite, Dolomite, Quartz content. The final fraction, S1.600 mainly contained Kaolinite, Dolomite, Calcite, and Quartz, heavier than 1.60 g/cm^3 media and reported as sink. SEM maps showed the raw coal had diverse particles with varying macerals and minerals.

Float-Sink concentrated vitrinite-rich particles in first fractions, and mineral content increased with the fraction density.

Microlithotype particle classification for the raw coal and its float-sink fractions is in **Fig. 3** and **Fig. 4**. The raw coal contained diverse microlithotypes, comprising 26.32% (by mass) Vitrite80, 21.52% Vitrite60, 14.95% VITSF, 13.59% FUS, and negligible amounts of other types. Tight density range float-sink processing resulted in significant enrichment of Vitrite80 in S1.275-F1.30 (94.70%) and S1.30-F1.325 (96.37%). The concentration of Vitrite80 particles declined with higher density fractions. S1.425-F1.45 fraction had 53.32% Vitrite60, an improvement. S1.525-F1.550 to S1.60 showed slight increases in Fusinite60 and FUS (5.76%, 12.65%). ALL peaked at 29.47% in S1.575-F1.60. CarboArg_MM20 and CarboArg_MM30 were 11.84% and 15.00% in S1.60. CarboPoly_MM20 increased to 10.58%. Adequate enrichment of Vitrite80 was achieved. There was some improvement in Vitrite60, FUS, and ALL particles, but these did not reach the 80% threshold. Further processing with narrower density ranges in 0.01 g/cm^3 increments is suggested for better separation of the Vitrite60, FUS and All particle types.

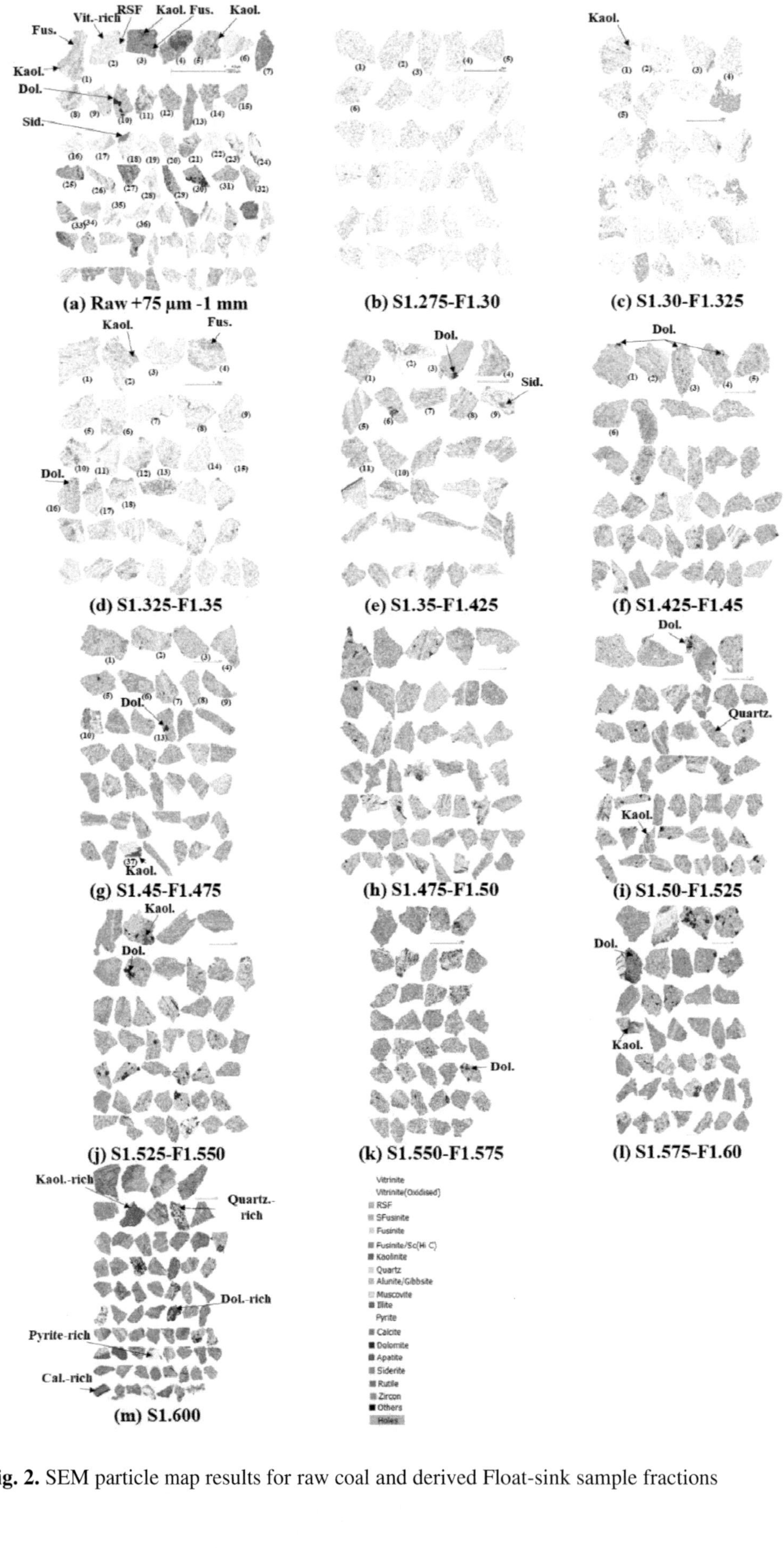

Fig. 2. SEM particle map results for raw coal and derived Float-sink sample fractions

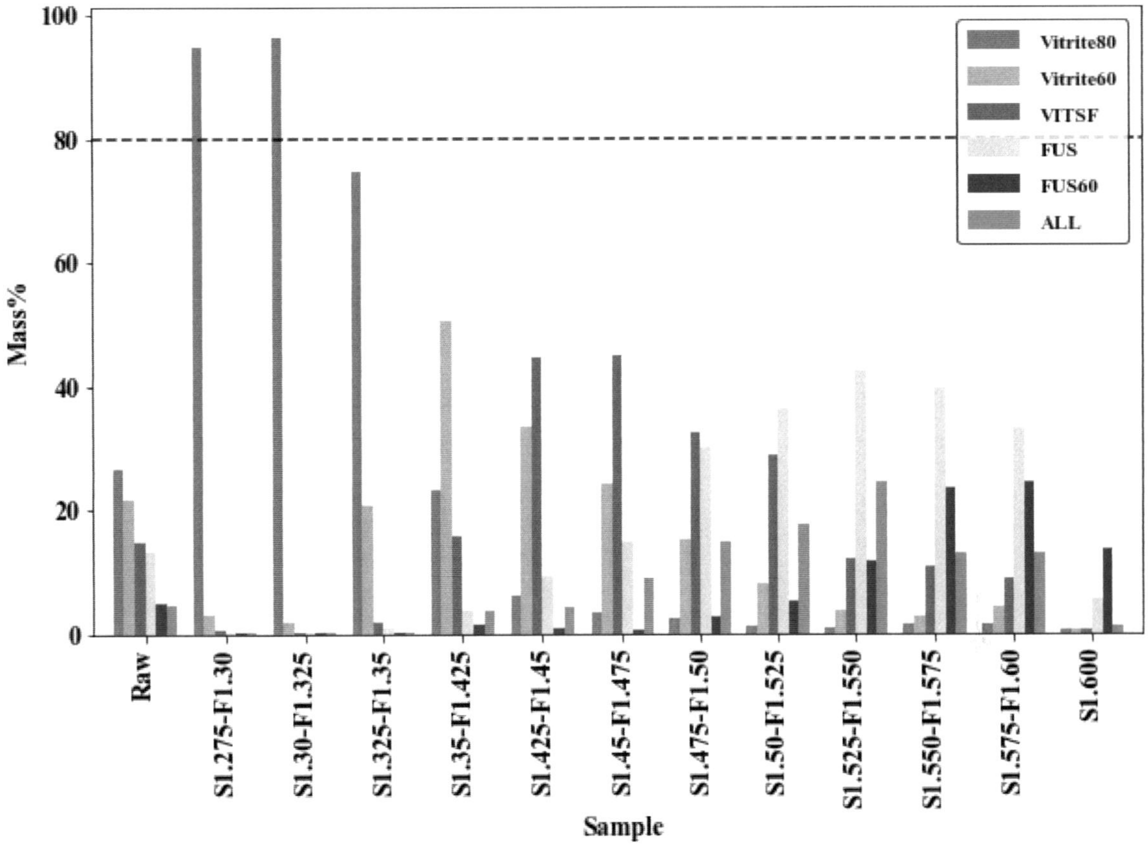

Fig. 3. SEM microlithotype particle type classification distribution results (maceralic group) for raw coal and derived Float-sink sample fractions

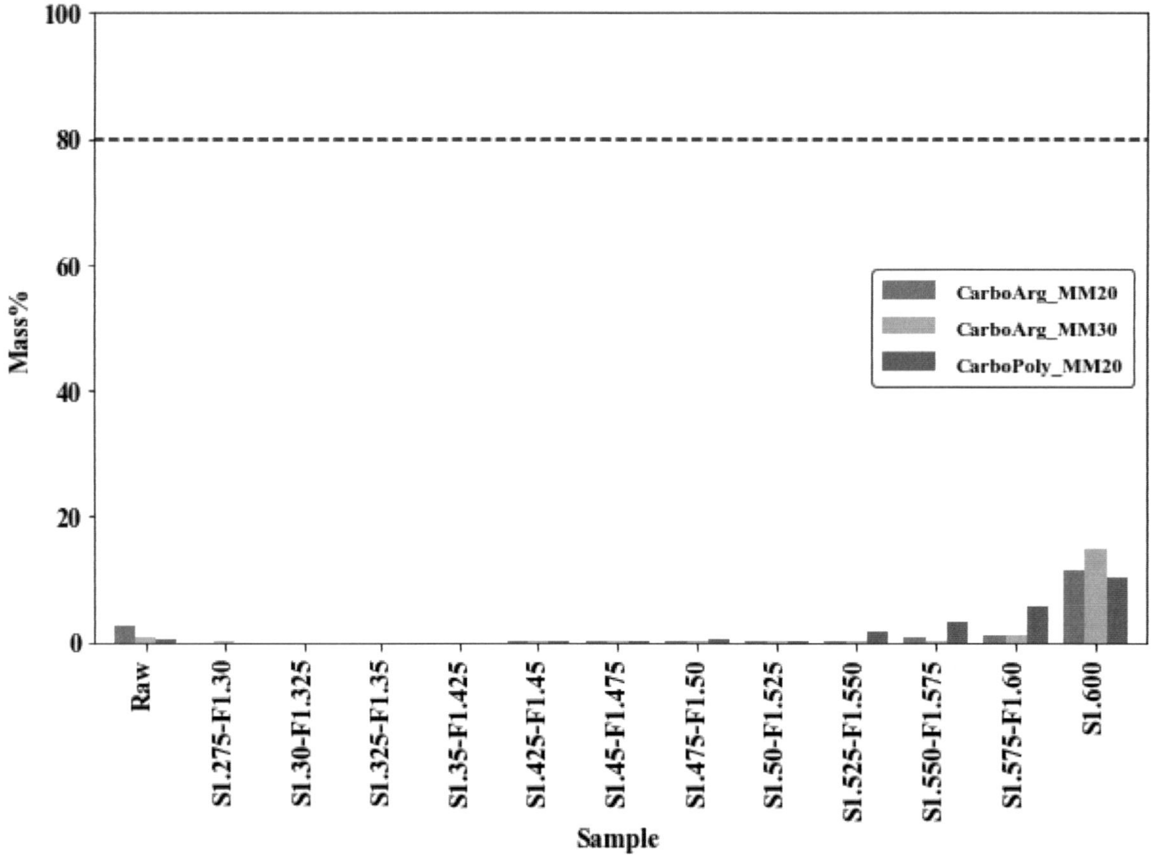

Fig. 4. SEM microlithotype particle type classification results (carbominerites group) for raw coal and derived Float-sink sample fractions

3.2. TGA ignition temperature results

The relationship between TGA ignition temperature and coal microlithotype particle type mass distribution results is illustrated in **Fig. 5**. Logarithmic feature transformation was implemented for the microlithotype mass concentrations (x-axis) to provide improved interpretability for the non-linear data. The analysis reveals a significant trend, as the concentration of Vitrite80 particles increases, the ignition temperature decreases. This trend is supported by a high R-squared value of 0.95, indicating a strong correlation between the ignition temperature and the proportion of Vitrite80 particles. This correlation can be attributed to the high content of reactive maceral, Vitrinite, within Vitrite80 particles. Vitrinite is known for its higher reactivity, which facilitates quicker ignition in comparison to less reactive macerals.

A similar decreasing trend in ignition temperature was observed for Vitrite60 particles, although with a slightly lower R-squared value, suggesting a weaker but still significant correlation.

Other particle types rich in inertinite macerals, such as FUS and FUS60, exhibited the opposite trend. As the concentration of these particle types increased, the ignition temperature also increased, indicating that inertinite macerals tend to resist ignition. The R-squared values for FUS and FUS60 were 0.75 and 0.67, respectively, indicating a moderate correlation. Similar behavior was observed in the All particle type category, where a comparable trend with an R-squared value of 0.65 was reported.

These results suggest that microlithotype particle types with high concentrations of reactive macerals, such as Vitrinite, contribute to lower ignition temperatures, while those rich in inertinite macerals exhibit higher ignition temperatures. These findings are consistent with the results of other studies on TGA combustion characteristics studies, particularly concerning the effect of inertinite content on ignition temperature. Rybak et al. [48] highlighted that increasing inertinite content, regardless of the coal's geological origin, leads to a rise in ignition temperature. This supports the current study's observation that coal particles rich in inertinite macerals exhibit higher ignition temperatures. Similarly, Behera & Nandi [49] and Milligan et al. [50] provide further validation of the finding. Behera explains that the increased inertinite content reduces reactivity, making coal ignition more challenging. Ryback et al. [48] noted that the measuring ignition temperature of coal particles using TGA and Differential Scanning Calorimetry (DSC) could be used to predict the ignition potential of any coal under conditions closest to those prevailing in pulverised fuel flames.

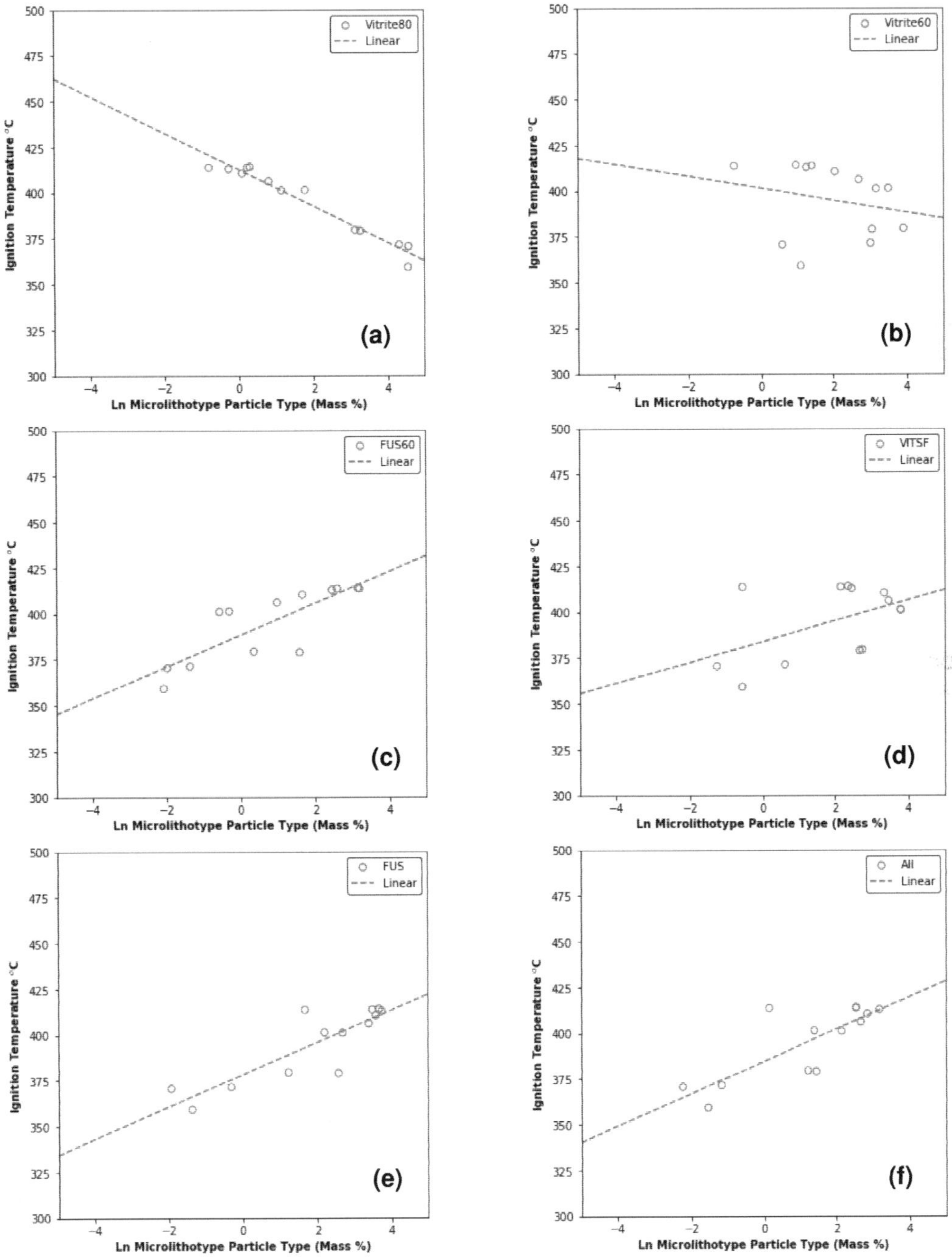

Fig. 5. TGA ignition temperature as a function of SEM coal microlithotype particle type mass distribution

3.3. TGA burnout temperature results

The correlation between TGA burnout temperature and coal microlithotype particle type mass distribution, is shown in **Fig. 6**. Logarithmic feature transformation was implemented for the microlithotype mass concentrations (x-axis) to provide improved interpretability for the non-linear data. The results indicate that as the concentration of Vitrite80 increases, the burnout temperature decreases. This can be attributed to the higher proportion of reactive macerals, particularly vitrinite, present in these particles, which are known for their enhanced reactivity and accelerate the combustion process, leading to a reduction in burnout temperature.

Conversely, particle types rich in inertinite, such as FUS60, FUS, and the All categories, display the opposite trend. These particles exhibit higher burnout temperatures, which can be linked to the lower reactivity of inertinite macerals.

These results suggest that microlithotype particle types with high concentrations of reactive macerals, such as Vitrinite, contribute to lower burnout temperatures. In comparison, those rich in inertinite macerals exhibit higher burnout temperatures. Similar findings have been consistently reported in various studies. Louw et al. [51] found that inertinite-rich char exhibited slightly higher burnout temperatures compared to vitrinite-rich char. This trend is supported by tests conducted by Morgan et al. [52], which demonstrated that burnout temperatures are lower for vitrinite-rich samples than for inertinite-rich ones. This behavior is attributed to the slower combustion rate of inertinite chars compared to vitrinite chars, as also noted by Jones et al. [53]. Behera (2021) further corroborated these findings by correlating coal inertinite content with burnout temperatures. The study revealed that an increase in the proportion of inertinite in coal results in slightly higher burnout temperatures, primarily due to the lower concentration of reactive macerals. Similarly, Milligan et al. [50] also reported that inertinite-rich coal samples tend to have higher burnout temperatures.

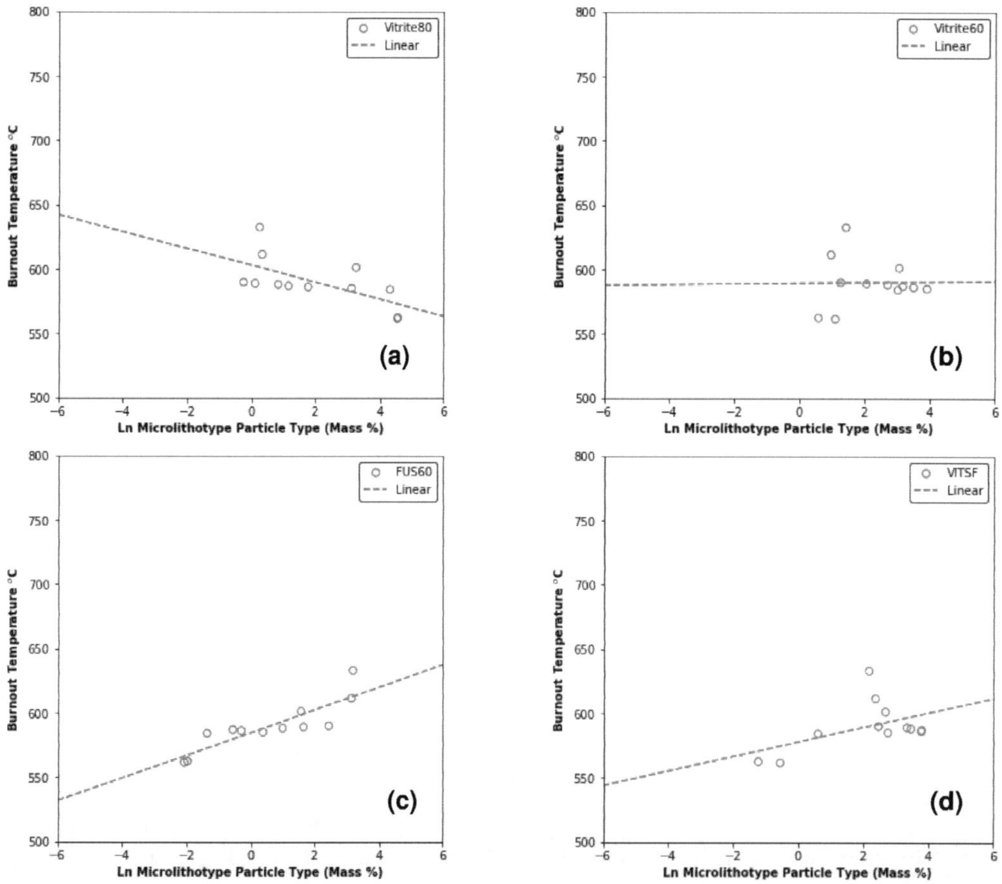

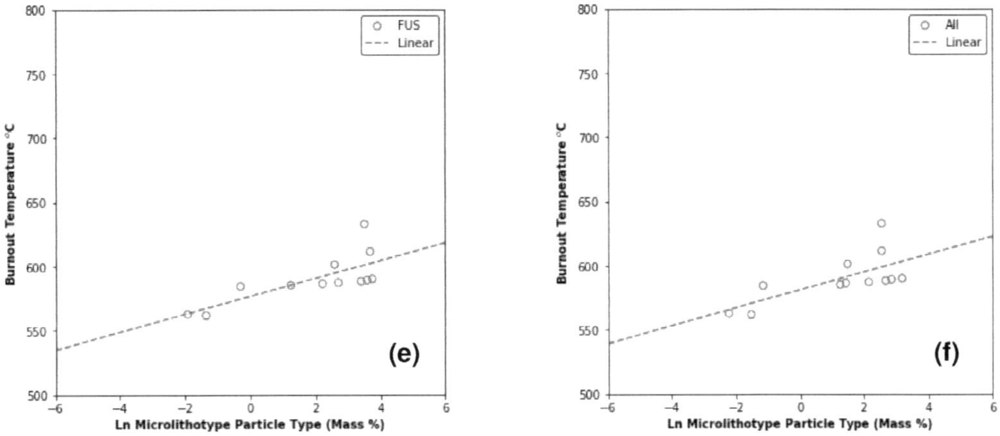

Fig. 6.TGA burnout temperature as function of SEM microlithotype particle type mass%

3.4. TGA peak temperature results

Fig. 7 illustrates the relationship between peak temperature and the proportions of coal microlithotype particle types. Logarithmic feature transformation was implemented for the microlithotype mass concentrations (x-axis) to provide improved interpretability for the non-linear data. The results indicate that an increase in the concentration of Vitrite80 particles correlates with a decrease in peak temperatures. This behavior can be attributed to the higher reactivity of the Vitrinite-rich Vitrite80 particles, which facilitates earlier and more rapid combustion, leading to lower peak temperatures. Conversely, particle types rich in inertinite macerals, such as FUS60, FUS, and VITSF, exhibit the opposite trend, where an increase in their concentration is associated with higher peak temperatures. This is likely due to the lower reactivity of inertinite-rich particles, which results in slower combustion and consequently higher temperatures at the combustion peak.

These findings are consistent with previous studies on the influence of maceral content on peak combustion temperatures. Morgan (1986) reported that vitrinite-rich samples exhibit lower peak temperatures compared to inertinite-rich samples. Similarly, Behera (2021) demonstrated that an increase in the proportion of inertinite in coal results in higher peak temperatures due to the lower reactivity of inertinite macerals. Milligan (1999) further corroborated these observations, noting that inertinite-rich samples produce higher peak temperatures during combustion.

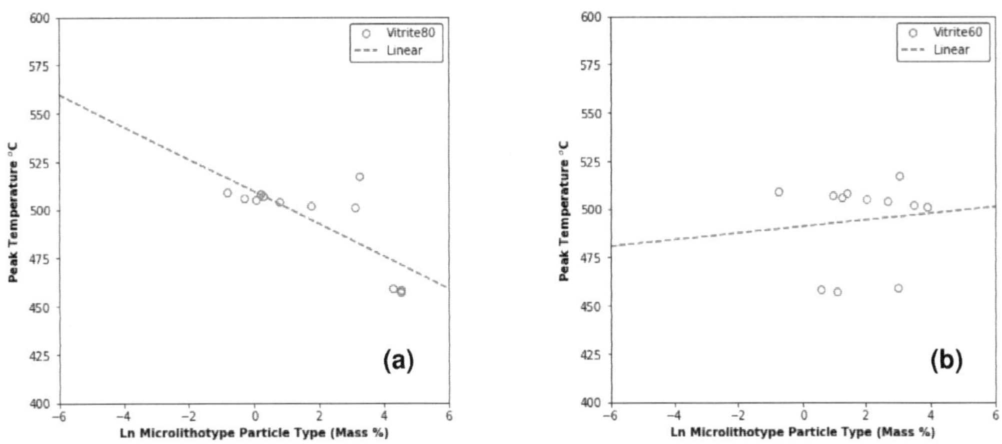

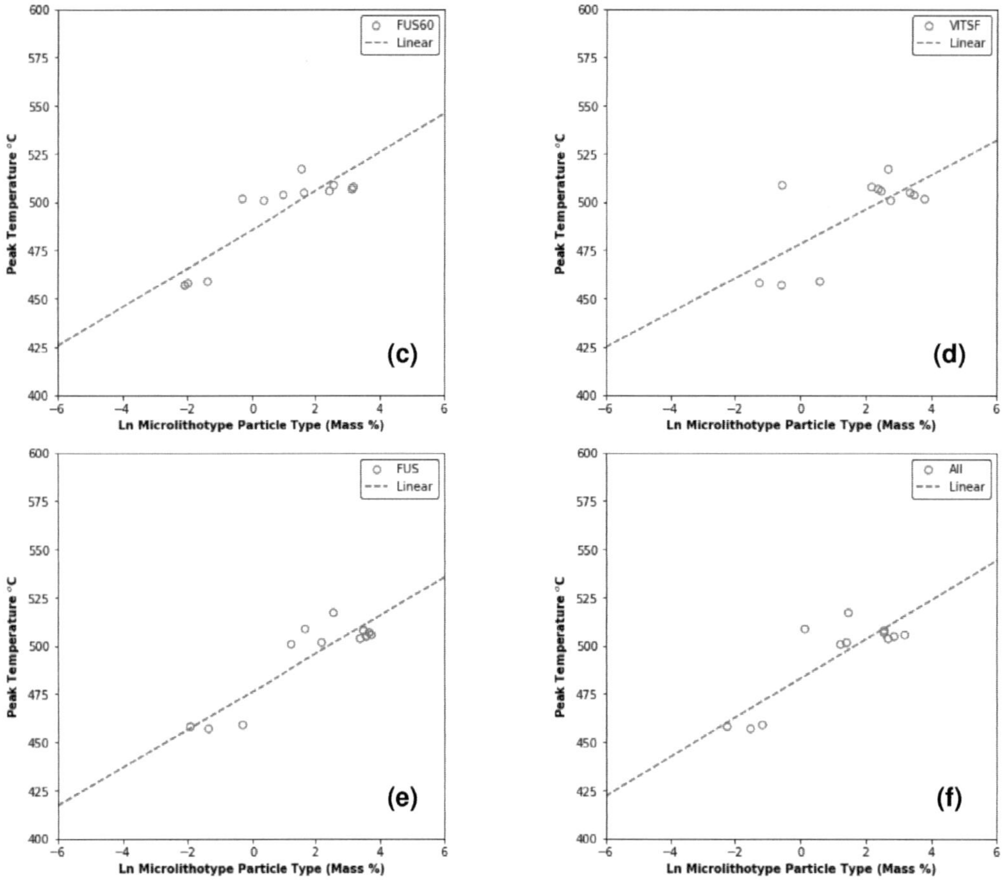

Fig. 7. TGA peak temperature as a function of SEM microlithotype particle type mass%

3.5. TGA reactivity results

The TGA reactivity results were correlated with coal microlithotype particle type concentration, as shown in **Fig. 8**. Logarithmic feature transformation was implemented for the microlithotype mass concentrations (x-axis) to provide improved interpretability for the non-linear data. An increase in Vitrite80 particles led to higher reactivity, which can be attributed to the dominance of vitrinite, a maceral known for its higher reactivity. Conversely, the results for FUS60, FUS, and All particle types demonstrated the opposite trend, with reactivity decreasing as the concentration of inertinite-rich particles increased.

These findings are consistent with other TGA combustion studies. Behera & Nandi [49] demonstrated that an increase in vitrinite content in coal leads to higher reactivity, which can be attributed to the higher proportion of reactive macerals in the sample. In contrast, inertinites, as noted by Milligan et al. [50], exhibit significantly lower reactivity because of their highly aromatic nature, which results in a more structured and stable carbon matrix that resists combustion. Thus, the maceral composition of coal plays a critical role in determining its reactivity, with vitrinite-rich samples being more reactive compared to inertinite-rich ones.

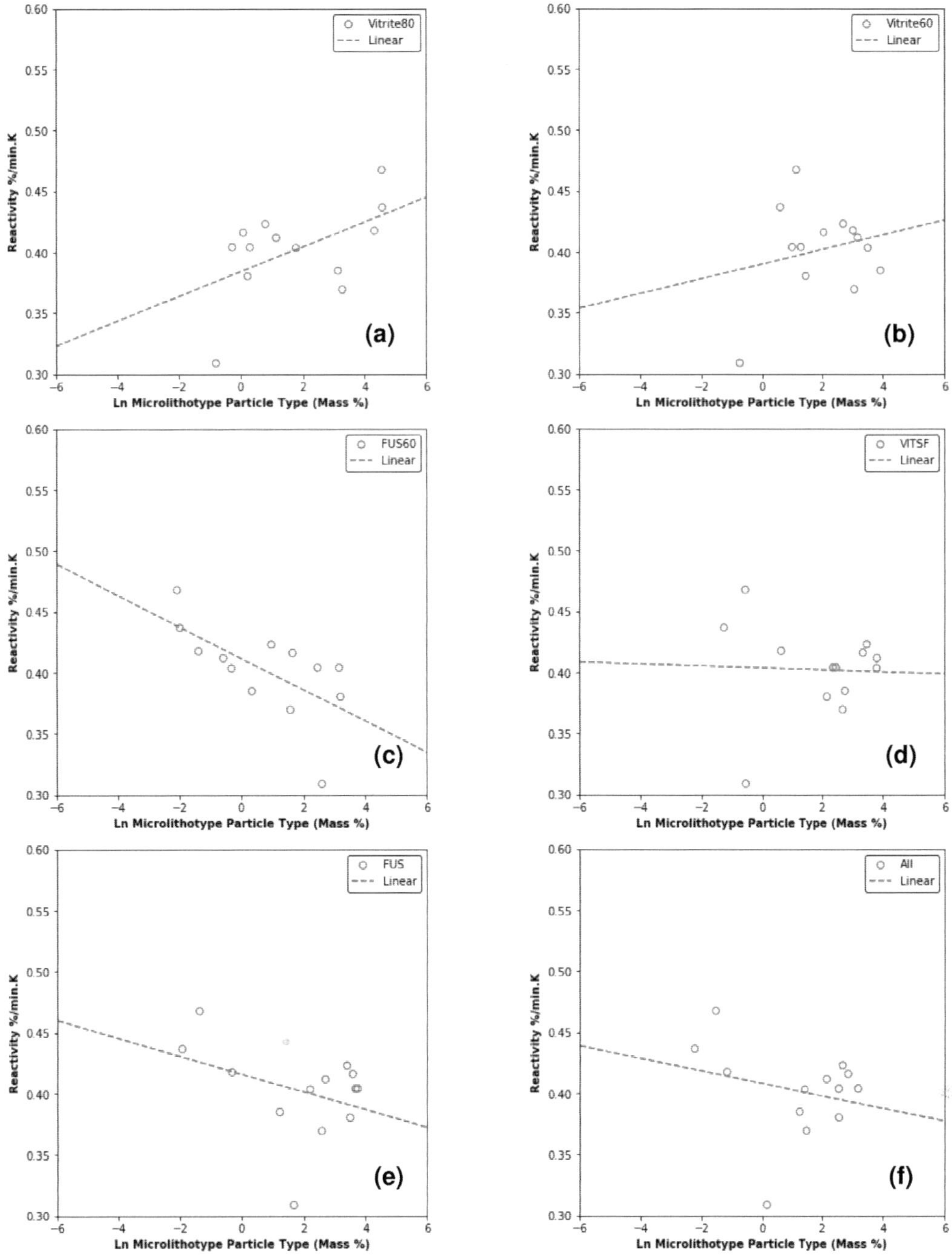

Fig. 8. TGA reactivity as function of SEM microlithotype particle type mass%

These results have demonstrated a clear relationship between the proportion of coal microlithotype particle types and TGA combustion characteristics, including ignition, burnout, peak temperatures, and reactivity. These findings are crucial for enhancing the accuracy of coal combustion models, as they highlight the importance of accounting for inter-particle variability. Incorporating these relationships into combustion models would improve the prediction of combustion behavior, particularly in coal blends.

4. Conclusions

The conclusion from the study on combustion characteristics of coal microlithotype particle types are as follows:

- The ignition temperature decreases with an increase in the proportion of Vitrite80 particles, which has been attributed to their high content of reactive Vitrinite maceral. A similar trend was observed for the results of the Vitrite60 particle type. Conversely, FUS and FUS60 particles that are rich in inertinite macerals exhibited the opposite trend.

- The sample burnout temperature decreased with increasing concentrations of Vitrite80 particles.
- The concentration of Vitrite80 particles influenced peak temperature and reactivity.

This study highlights the significance of microlithotype particle type particle proportion to achieve lower ignition and burnout temperatures, ultimately enhancing coal reactivity. These are important characteristics of optimal boiler performance.

5. Recommendations

The recommendations of the study are as follows:

- Further research should be conducted to determine the interactions between various microlithotype particle types during combustion.
- The combustion characteristics of chars derived from various coal microlithotype particle types should be investigated.
- Coal microlithotype particle-type characteristics should be incorporated into combustion models.

Acknowledgements

The authors acknowledge the financial support from Eskom Research, Testing, and Development and the support provided by the Eskom Power Plant Engineering Institute (EPPEI). They extend their gratitude to individuals who contributed significantly to this study: Bonny Nyangwa, Dr Chris Van Alphen and Priven Rajoo for technical guidance, Shinelka Singh and Thulisile Sishuba for support with QEMSCAN and TESCAN analyses, Prof Rosemary Falcon, Dr Samson Bada, and Dr Mark Van der Riet for advice during PhD meetings, Patrick Musie, Phumzile Mabidi, Xolani Ngubeni, and Olifant Sinthumule for chemical analysis aid, Mandla and John from Siza Laboratories for sample preparation, Sean Harbinson and colleagues at Bureau Veritas Laboratories for float-sink processing advice, Boitshoko Mashike, Bongani Khumalo, Molly Minyuku, Rendani Tshotheli and Prof Schmitz for export coal qualities assistance, and Jeffrey Mesolo for sample collection aid, Olifant Sinthumule and Prof Krzysztof Czajka for assisting with TGA tests and providing advice.

References

[1] X. Prevost, "Current and future use of coal in South Africa," in *Fossil Fuel Foundation Energy Conference*, 2015.

[2] DMRE, *Department of Mineral Resources and Energy Intergrated Resource Plan IRP2019,* 2019.

[3] DMRE, *Integrated Resource Plan 2023,* 2024.

[4] IEA, "Coal 2022 Analysis and forecast to 2022," 2022.

[5] I. Suarez-Ruiz and C. R. Ward, Applied coal petrology, 2008.

[6] M. Jakob, J. C. Steckel, F. Jotzo, B. K. Sovacool, L. Cornelsen, R. Chandra, O. Edenhofer, C. Holden, A. Lochel, T. Nace, N. Robins, J. Suedekum and J. Urpelainen, "The future of coal in a carbon-constrained climate," 2020.

[7] S. Ivanova, A. Vesinina, N. Fotina and A. Prosekov, "An overview of carbon footprint of coal mining to curtail greenhouse gas emissions," 2022.

[8] L. Moodley and P. Rajoo, "Duvha Power Station Unit 1 Corrosion Investigation," 2012.

[9] M. Blenkinsop, G. Twala, P. Gildenhuys and W. Van Remoortere, "Matimba Lower Seams Phase 3," 1994.

[10] D. Fong and H.-C. Magel, "Matimba Furnace Simulation," 1997.

[11] P. Rajoo, "Matla Seam 2 and 4 coal tests at the Pilot Scale Combution Test Facility," 2002.

[12] L. Bartonova, "Unburned carbon from coal combustion ash: An overview," 2015.

[13] M. Torresi, B. Fortunato, S. M. Camporeale and A. Saponaro, "CFD Modeling of pulverised coal combustion in an industrial burner," 2012.

[14] J. Pallares, I. Arauzo and A. Williams, "integration of CFD codes and advanced combustion models for quantitative burnout determination," 2007.

[15] R. Vuthaluru and H. B. Vuthaluru, "Modelling of a wall fired furnace for differenct operating conditions using Fluent," 2006.

[16] M. Xu, J. Azevedo and M. G. Carvalho, "Modelling of front wall fired utility boiler for different operating conditiosn," 2001.

[17] Z. F. Tian, P. J. Witt, M. P. Schwarz and W. Yang, "Numerical modelling of pulverised coal combustion," in *Handbook of multiphase flow science and technology*, 2023, p. 339.

[18] M. B. Gandi, R. Vuthaluru, H. Vuthaluru, D. French and K. Shah, "CFD based predition of erosion rate in large scale wall-fired boiler," 2012.

[19] C. Sheng, B. Moghtaderi, R. Gupta and T. F. Wall, "A computational fluid dynamics based study of the combustion characteristics of coal blends in pulverised coal-fired furnace," 2004.

[20] A. Williams, R. Backreedy, R. Habib, J. Jones and M. Pourkashanian, "Modelling coal combustion: The current position," 2002.

[21] L. Moodley, "Lethabo Power Station Units 2 & 3 combustion tests," 2013.

[22] J. C. Crelling, E. J. Hippo and B. A. Woerner, "Combustion characteristics of selected whole coal and macerals," 1992.

[23] N. Soundarrajan, N. Krishnamurthy and S. V. Pisupati, "Physical and Chemical Characterization of Coal Particles Used as Entrained Flow Gasifier Feedstock: Heterogeneity in Mineral Matter Distribution," 2012.

[24] G. Marangwanda, D. Madyira and T. Babarinde, "Coal Combustion models: A revew," 2019.

[25] R. Falcon and A. Ham, "The characteristics of South African coals," 1988.

[26] P. Rajoo, "The application of pilot scale coal evaluation to full scale boilers," 2010.

[27] S. Singh, "The potential of QEMSCAN in predicting washability of Mpumalanga coals," 2015.

[28] J. Yu, J. A. Lucas and T. F. Wall, "Formation of the struture of chars during devolatilisation of pulverised coal and its thermoproperties A review," 2007.

[29] ICCP, "The new inertinite classification (ICCP System 1994)," 2001.

[30] R. H. Hurt and R. E. Mitchell, "On the combustion kinetics of heterogeneous char particle populations," 1992.

[31] J. Gibbins, T. Beeley, J. Crelling, A. Scott, N. Skorupska and J. Williamson, "Observations of heterogeneity large pulverised coal particles," 1999.

[32] C. van Alphen, Interviewee, *QEMSCAN coal particle type characterisation.* [Interview]. 4 April 2019.

[33] G. H. Taylor, M. Teichmuller, A. Davis, C. F. K. Diessel, R. Littke and P. Robert, Organic petrology, 1998.

[34] C. van Alphen, *TESCAN coal microlithotype analysis setup,* 2021.

[35] R. Backreedy, L. Fletcher, L. Ma, M. Pourkashanian and A. Williams, "Modelling pulverised coal combustion using detailed coal combustion model," 2006.

[36] K. P. Galvin, "Options for washability analysis of coal - A literature review," 2006.

[37] K. Bergh, "Trace element concentration reductin by benefication of Witbank Coalfield," 2010.

[38] R. M. S. Falcon and C. P. Synman, "An introduction to coal petrography: Atlas of Petrographic Consituents in the Bituminous coal of Southern Africa," 1986.

[39] S. Harbinson, *Float-Sink Test Methodology,* 2020.

[40] L. Moodley, F. Conradie, W. Schmitz, J. Mathews, H. Neomagus, C. Van Alphen and B. Nyangwa, "Microlithotype particle type classification and enrichment of coal macedral concentrates with Relfux Classfier, Float-Sink and RhoVol," 2024.

[41] Y. Lui, "Mineral matter organic matter association characterisation by QEMSCAN and application in coal utilisation," 2005.

[42] French, David; Ward, Colin R;, "The application of advanced mineralogical techniques to coal combustion product characterisation," 2009.

[43] S. Bada, Interviewee, *TGA microlithotype defined particle tests.* [Interview]. 04 December 2020.

[44] S. O. Bada, R. M. S. Falcon and L. M. Falcon, "characterisation and co-firing potential of a high ash coal with Bambusa balcooa," 2015.

[45] J. Chao, H. Yang, Y. Chen, J. Luv, H. Zhang and Q. Lui, "The investigation of the coal ignition temperature in oxygen-enriched atmosphere by TGA," 2013.

[46] J.-J. Lu and W.-H. Chen, "Investigation on the ignition and burnout temperature of bamboo and sugarcane bagasse by thermogravimetric analysis," 2015.

[47] K. Isaac, "The co-combustion performance of South African coal and refuse derived fuel," 2019.

[48] W. Rybak, W. Moron and J. Wach, "Ignition studies of high-vitrinite and high-inertinite coals using TGA DSC, DTIF, EFR and 20 L dust explosive chamber," 2021.

[49] D. Behera and B. K. Nandi, "Effect of coal particle density on coal properties and combustion characteristics," 2021.

[50] J. B. Milligan, K. M. Thomas and J. C. Crelling, "Temperature programmed combustion studies of coal and maceral group concentrates," 1997.

[51] E. B. Louw, G. D. Mitchell, J. Wang, R. E. Winans and J. R. Mathews, "Constitution of Drop-Tube-Generated Coal Chars from Vitrinite-rich and Inertinite-rich South African," 2015.

[52] P. A. Morgan, S. D. Robertson and J. F. Unsworth, "Combustion studies by thermogravimetric analysis," 1986.

[53] R. Jones, "Maceral and rank influences on the morphology of coal char," 1985.

2024 International Pittsburgh Energy and Carbon Management Conference

October 7 - 9, 2024

Abstract Submission

PROGRAM TOPIC:

COMBUSTION TECHNOLOGIES

Exact Title of Paper:

DROP TUBE FURNACE COMBUSTION CHARACTERISTICS AND COMPUTATIONAL FLUID DYNAMICS MODELLING OF COAL CHAR MICROLITHOTYPE PARTICLE TYPES

Presenting Author
Lesigen Moodley, Eskom Research, Testing and Development, Lower Germiston Road, Rosherville, Johannesburg, South Africa,
moodll@eskom.co.za

Co-Authors
Frikkie Conradie, Dr, North-West University, School of Chemical and Mineral Engineering, Potchefstroom, South Africa, frikkie.conradie@nwu.ac.za

Walter Schmitz, Prof, University of Witwatersrand, School of Mechanical, Industrial and Aeronautical Engineering, Johannesburg, South Africa, walter.schmitz@wits.ac.za

Hein Neomagus, Prof, North-West University, School of Chemical and Mineral Engineering, Potchefstroom, South Africa, hein.neomagus@nwu.ac.za

Bonny Nyangwa, Eskom Research, Testing and Development, Lower Germiston Road, Rosherville, Johannesburg, South Africa, nyangwtb@eskom.co.za

Kganuwi Kekana, Eskom Research, Testing and Development, Lower Germiston Road, Rosherville, Johannesburg, South Africa, kekanakg@eskom.co.za

Abstract:

This study investigates the char combustion characteristics of coal microlithotype particle type concentrates and their integration into Computational Fluid Dynamics (CFD) combustion modelling to simulate microlithotype particle type blends. The model predictions were validated against experimental measurements. Coal samples, enriched in various microlithotypes including Vitrite80, Vitrite60, VITRSF, VITSF and FUS, were pulverised and pyrolyzed in Eskom's Drop Tube Furnace (DTF) in a nitrogen atmosphere at 1400 °C to produce the char samples. These char samples were further pulverised and sieved to produce a size fraction of +38 -75 µm and combusted at wall temperatures 1000 °C, 1200 °C and 1400 °C. Solid residue samples were collected at different probe positions corresponding to varying particle residence times and analysed for ash yield at RT&D's Coal laboratory to determine the combustion efficiency at each probe position using the ash tracer method. A CFD Surface Reaction Discrete Phase Model (DPM) of the DTF was developed using Ansys Fluent software, incorporating a User-Defined Function (UDF) to set the char kinetic parameters for different microlithotype particle types. The results indicated the chars derived from the Vitrite80 rich coal samples were more reactive, ignited quicker and had a shorter burnout residence time than the other samples. Furthermore, the CFD modelling demonstrated that char burnout characteristics of the whole coal sample can be predicted by simulating its individual coal microlithotype particle type constituents. This study has demonstrated the potential of using CFD combustion modelling, including microlithotype particle type features, as an alternative to costly and time-consuming DTF experimental tests for evaluating the combustion characteristics of new coal sources.

1. Introduction

Coal is an important energy source for power generation [1, 2, 3, 4]. However, increasing environmental concerns surround coal mining operations and the associated emissions released during coal combustion [5, 6, 7]. Additionally, declining coal quality at power stations [1, 8] which has contributed to boiler combustion related issues such as primary ignition problems [9] slagging [10, 11], fireside corrosion [8], temperature excursions [10] and high unburnt carbon losses [12]. To ensure coal's continued use in power generation, it is crucial to minimise its environmental impact and address these operational challenges. This requires a comprehensive understanding of the combustion characteristics of new coal sources before they are fired in industrial boilers.

Computational Fluid Dynamics (CFD) simulations are widely used to assess combustion characteristics of new coal sources and blends [13], optimise the efficiency of energy systems [14], simulate changes to operating parameters [15, 16] and burner modifications [17]. Additionally, CFD models can track coal particle interactions [18] and simulate flow patterns within the boiler and model key combustion characteristics such as coal ignition, burnout, particle residence time and also simulate gaseous species formation e.g. Oxides of Nitrogen (NO_x) [19] and Sulphur Oxides (SO_x) and provide insights into temperature distribution and hot spots to minimize slagging formation. By integrating chemical kinetics, CFD allows for the simulation of complex chemical reactions during combustion, enabling the testing of different scenarios without the need for costly physical experiments. This allows for the identification of potential problems early and the testing of potential solutions prior to firing and implementing them in industrial boilers. However, while CFD has demonstrated accuracy in predicting the flow field [20] and gas temperatures during coal combustion, other predicted quantities such as gaseous emissions and burnout levels have shown to be less accurate [21]. Burnout predictions are particularly important since they are utilised to estimate the boiler furnace height during the design of new units and to evaluate the potential use of new coal sources for existing boilers. These predictions' inaccuracies may stem from the complex nature of coal and the inherent challenges in modelling its combustion behaviour [22, 23, 24]. Therefore, although CFD remains a valuable tool in coal combustion studies, further refinement of certain model outputs, particularly emissions and burnout, is necessary to improve prediction reliability.

Coal combustion behaviour is influenced by both external and inherent factors, which need to be included in Computational Fluid Dynamics (CFD) models for reliable predictions [25]. External factors, such as particle size, air temperature, burner swirl setting, design and furnace height, have been studied extensively [26]. In contrast, inherent factors, including the maceral and mineral content of coal, its rank, porosity, reactivity, exposed surface area, particle size, and degree of weathering, are equally significant but more challenging to model [25]. The heterogeneous nature of coal, especially after pulverization, introduces substantial variability in particle composition [27], which must be accounted for to ensure accurate combustion models. Macerals, the organic constituents of coal, influence physical and chemical properties, including reactivity and burnout behaviour, while mineral matter contributes to slagging. Pulverized coal samples are particularly heterogeneous, containing particles with vastly different maceral compositions, which behave differently upon heating. For instance, variations in maceral constituents can affect the yield of volatile matter, char reactivity, and ash chemistry [28, 29, 30], leading to different combustion outcomes for coal particles within the same sample.

Gibbins et al. [31] noted that conventional coal analyses are not well-suited to describe this heterogeneity. Traditional methods assume coal is homogeneous and are focused on obtaining average values, which may not reflect the true behaviour of a heterogeneous particle population. An alternative method based on maceral associations should be used to describe heterogeneity in pulverised coal particles. Average coal properties are appropriate for models such as overall plant Mass and Energy Balances (MEB). Additionally, the combustion behaviour of a heterogeneous array of coal particles can significantly differ from that of a homogeneous sample with the same average composition. In particular, individual particle properties are crucial in gas-solid char combustion, influencing the degree of unburnt carbon leaving the boiler. Therefore, understanding and incorporating coal heterogeneity into combustion models, particularly CFD simulations, is essential for accurately predicting ignition, burnout, and emissions under industrial conditions.

Eskom has undertaken extensive research to understand the heterogeneity of pulverized coal particles and their impact on power station performance, particularly in relation to slag formation, high unburnt carbon, flame stability, and elevated furnace exit gas temperatures. This research led to the development of a pioneering approach by Eskom's Research, Testing, and Development (RT&D) department, utilizing **Q**uantitative **E**valuation of **M**inerals by **Scan**ning Electron Microscopy (QEMSCAN) to analyze coal particles [32]. QEMSCAN produces detailed false-color particle maps, distinguishing between homogeneous particles, such as those composed solely of macerals e.g. Vitrinite or minerals e.g. Quartz, Kaolinite, and heterogeneous particles containing various maceral and mineral associations in differing proportions. Furthermore, to build on the Scanning Electron Microscopy (SEM) analysis, RT&D introduced the Microlithotype Particle Classification System (MPCS), which comprises 49 distinct particle types [32, 27]. This classification expands on the traditional classification used in petrographic analysis [33] by grouping coal particles into three categories: maceralic, carbominerites, and minerites. The maceralic group includes particles rich in macerals, such as Vitrite80, containing ≥80% (by mass) Vitrinite, or mixed particles like VITRSF, which contains ≥60% Vitrinite and Reactive Semifusinite. The carbominerites group encompasses coal-mineral combinations, such as carbosilicate_MM40, which contains coal and quartz ≥40%. Finally, the minerites group is composed primarily of minerals, like Pyrite Cleat, which contains ≥60% Pyrite. This advanced particle classification system allows RT&D to quantify the distribution of microlithotype particle masses in pulverized coal samples from South Africa and internationally, revealing significant differences in their mass distributions [32, 27]. In addition, RT&D has adapted a **TES**la **SCAN**ning (TESCAN) Scanning Electron Microscope to quantify microlithotype particle distribution [34]. While this work has significantly advanced the characterization of coal particle heterogeneity, further research is needed to explore the char combustion characteristics and interactions of these coal microlithotype particle types, which remain poorly understood.

Previous CFD studies applied a maceral correction factor to improve the accuracy of char burnout models by adjusting for the influence of maceral content on reactivity [35, 14]. However, these studies primarily focused on average properties, simulating coal particles with similar maceral compositions and, consequently, similar char reactivity. While the significance of coal inter-particle heterogeneity is well recognized, its effects have not yet been integrated into CFD combustion models. This omission limits the ability of current models to make reliable predictions regarding combustion behaviour. Capturing this heterogeneity is essential for improving the accuracy of combustion predictions, particularly in relation to ignition, burnout, and slagging phenomena. An understanding of the heterogeneous nature of pulverized coals is fundamental to the development of more precise conceptual and quantitative models for coal processing and combustion [31]. By incorporating inter-particle heterogeneity into CFD models, researchers can better simulate the complex interactions between maceral and mineral constituents, ultimately leading to improved plant optimization, enhanced blending strategies, and more effective environmental mitigation efforts. Therefore, further research is necessary to determine the char combustion characteristics of coal microlithotype particle types, including inter-particle maceral and mineral heterogeneity, and to integrate these features into CFD models for more accurate and reliable combustion simulations.

This study investigates the DTF char combustion characteristics of coal microlithotype particle type concentrates and their integration into CFD combustion modelling to simulate microlithotype particle type blends.

2. Materials and Methods

2.1. Sample origin and coal processing

Coal A, a South African export coal sample that is vitrinite-rich and low-ash yield from the Witbank Coalfield was selected for this study. This sample was chosen because of its lower ash yield as compared to local power station coals, making it more efficient to concentrate microlithotype particle types. The bulk coal sample was crushed to a size of -1 mm. Float-Sink separation of the coal sample using tight density ranges was conducted at Bureau Veritas Laboratories in Centurion. The dense liquid for separation (1.325, 1.350 and 1.450 g/cm³) was prepared by blending perchloroethylene (1.62 g/cm³) with white spirits (0.79 g/cm³). Liquid density was verified using a pycnometer or a calibrated measuring cylinder, and a calibrated scale. Heavy liquids such as zinc chloride were avoided to prevent pore adhesion, which could interfere with subsequent coal analysis [36, 37]. Based on previous coal processing tests, a liquid density of 1.325 g/cm³ was selected to produce a high concentration of Vitrite80 particles while densities 1.350 and 1.450 g/cm^3 were selected to enriched concentrations of Vitrite60 and VITSF microlithotype particle types respectively [38]. A 500 g coal sample was mixed with the dense liquid in a 25-liter container and manually agitated for 45 minutes. Next, the mixture was poured into a separation funnel and allowed to settle for up to 24 hours. The float fraction was collected using filter paper, rinsed, and air-dried. The sink fraction underwent further processing to yield the final fractions [39]. The Float-Sink processing produced three sample fractions. Additional procedural details on the Float-Sink separation method are outlined in Moodley et al. [38].

Coal B, a Polish coal sample from the Kraków region that is vitrinite-rich, low-ash yield was selected for this study and underwent coal processing using North-West University's laboratory Reflux Classifier [40]. The equipment was operated in batch mode with an initial 2 kg coal sample. Firstly, the Reflux Classifier was flushed to remove any particles that might be in the bed zone, lamella, overflow box and outlet pipe. Next, the coal sample was added through the feed port. The pump was started, and the equipment was set to the desired inlet water flow rate of 1500 L/hour and allowed to stabilise. Then, the overflow product fractions were collected in a 100 litre plastic drum and the excess water was decanted. These samples were oven dried overnight at 40 °C.

2.2. SEM analysis and coal microlithotype particle type classification

SEM sample preparation involved subdividing each coal sample into ~0.5 g portions using Quantachrome's micro riffle. Each portion was mixed with molten carnauba wax and moulded into 30 mm diameter blocks [41, 27]. Carnauba wax was chosen because it allows for the analysis of both organic and inorganic constituents, thereby facilitating clear differentiation between coal particles from the mounting medium. The samples were polished and carbon coated to reduce surface roughness and charge issues [42]. SEM analysis was conducted using Eskom RT&D's TESCAN TIMA instrument. During the analysis, the Backscattered Electrons (BSE) and Energy Dispersive X-Rays (EDS) produced were compared with the Species Identification Protocol (SIP) file for phase identification. Once all points were analysed on individual coal particles, the system generated false-colour particle maps which illustrated the maceral and mineral distribution [32]. Verification SEM results involved a comparison of total mineral matter content with proximate ash yield.

TIMA software was utilised to classify coal particles in accordance with Eskom RT&D's Microlithotype Particle Classification System (MPCS), following the novel methodology developed by Van Alphen [34]. The SEM data, including particle maps and microlithotype particle type distribution data, were exported to Excel format for subsequent processing. This allowed for the annotation of particle maps and the creation of graphical representations that illustrated the distribution of microlithotype particle types.

2.3. DTF combustion characteristics of chars derived from coal microlithotype particle type concentrates

Coals A and B along, with their derived coal microlithotype concentrates, were pulverised and sieved to -150 μm and pyrolysed in a nitrogen atmosphere at 1400 °C for a period of 2 seconds in Eskom's DTF to generate the char samples. These samples were further pulverised and sieved to produced two size fractions: -38 μm and +38-75 μm. The +38-75 μm char fraction which is expected to have a longer burnout time, was selected for this study and was submitted for proximate, ultimate and gross calorific value analyses. These samples were evaluated in the DTF under simulated full-scale combustion conditions. These tests were performed at three furnace wall temperatures 1000 °C, 1200 °C and 1400 °C. During each test samples were extracted at different probe depths corresponding to varying particle residence times and the furnace exit flue gas oxygen (O_2) concentration was maintained at 3.0% (volume). The extracted samples were analysed to determine their ash yield and thereafter char combustion efficiency (conversion) was calculated at each probe depth using the Ash Tracer Method. These results were compared to assess any relationships between conversion and the concentration of microlithotype particle types.

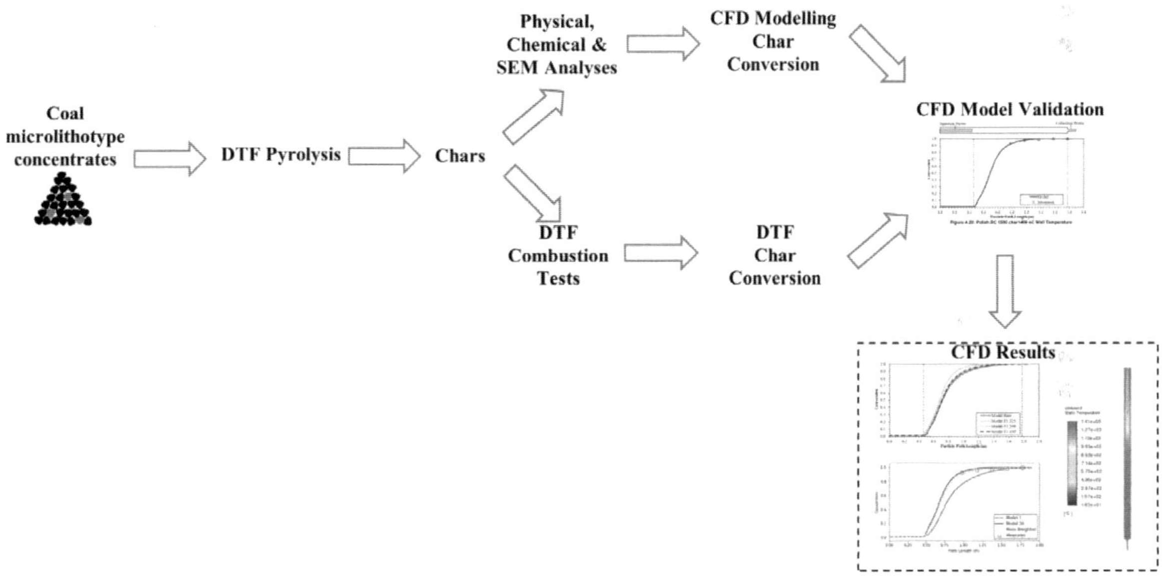

Fig. 1. Coal microlithotype particle type DTF char combustion test and CFD modelling process

2.4. CFD char combustion model for simulating coal microlithotype particle type concentrates and blends

A CFD combustion model of Eskom's DTF was developed using the Ansys Fluent 2023 R1 software. The CFD model solves equations for both fluid (continuous) and solid (discrete) phases. Ansys Fluent calculates the fluid phases by solving the governing conservation equations for mass (continuity), momentum, energy (heat transfer) and species transport (mixing or reactions). This study uses the Multiple Surface Reaction Model which assumes the surface species becomes particle surface species and could be depleted or produced [43]. The solid particles are modelled as spheres dispersed in the continuous phases, this allows the solid phase to be modelled as reacting particle and its effect on the continuous phase to be simulated. The combustion particle type enables the simulation of heating, the evolution of volatiles, swelling, and heterogeneous surface reactions. Discreate

Phase Model (DPM) was used to simulate the solid phase. The experimental data including the primary and secondary air velocities, their species volume fraction and temperature, the combustion chamber wall temperature were specified as Boundary Conditions in the model. Two User Defined Functions (UDFs) were used to implement the variable DTF wall temperature and the Surface Reaction Model parameters. The CFD char combustion model uses different char kinetic parameters for the various microlithotype particle types. Depending on the coal char particle number (#) being solved, the model selects the corresponding char kinetic parameters (activation energy and pre-exponential factor). Thereafter, the particle energy balance is solved until the model converges then the data is stored. DPM variable results including particle residence time, temperature, mass, velocity and diameter were extracted from the model during post processing. The particle results were used to calculate the predicted char conversion and compared graphically with the experimental data at each probe position.

3. Results & Discussions

3.1. Scanning Electron Microscopy results

SEM results were validated by comparing the total mineral matter content with proximate analysis ash yield, revealing strong agreement and confirming the SEM reliability. TESCAN false colour particle map results details the maceral and mineral distribution and associations within individual coal particles. Coal A raw and the derived Float-Sink fractions particle map results are shown in **Fig. 2 (a) to (d)**. Coal A raw feed **Error! Reference source not found. (a)**) contained particles comprising of vitrinite associated with fusinite and kaolinite (Particles 1, 8 and 11). Other particles were vitrinite-rich with finely distributed RSF inclusions (Particles 2, 6 and 9) or fusinite associated with kaolinite (Particles 3,4 and 7). Some contained vitrinite associated with fine inclusions of semifusinite and kaolinite (Particle 5). There are particles that contain high proportion of kaolinite (Particle 7), vitrinite with dolomite inclusions (Particle 10), vitrinite mixed with fusinite, semifusinite and RSF (Particle 12 and 21) and fusinite associated with vitrinite, semifusinite and kaolinite (Particle 13). Certain particles mainly contain minerals e.g. pyrite (Particle 16). These particle maps have shown the complex associations of maceral and minerals found in Coal A raw particles.

The first fraction obtained from the float-sink tests was the F1.325 (**Fig. 2 (b)**) and most of the particles are similar containing a high proportion of vitrinite with fine inclusions (Particles 2 to 5). The next fraction acquired from the float-sink processing was F1.350 (**Fig. 2 (c)**) which contains particles that are vitrinite-rich and associated with small quantities of fusinite, kaolinite and dolomite (Particles 2, 4 and 16). The proportion of kaolinite, fusinite and dolomite in the particles has increased slightly compared to the previous fraction. Next the F1.450 fraction was obtained (**Fig. 2 (d)**) and contained particles that are rich in vitrinite that are associated with fusinite and quartz/kaolinite (Particles 1, 2 and 7). Other particles in this fraction contain vitrinite mixed with fusinite containing dolomite and kaolinite inclusions (Particle 3), some vitrinite mixed with semifusinite (Particle 5), vitrinite mixed with fusinite and semifusinite with small quantities of kaolinite (Particle 4, 6, 8) while some are vitrinite-rich with an included siderite grain (Particle 9). The F1.450 fraction contains more mineral matter than previous fractions.

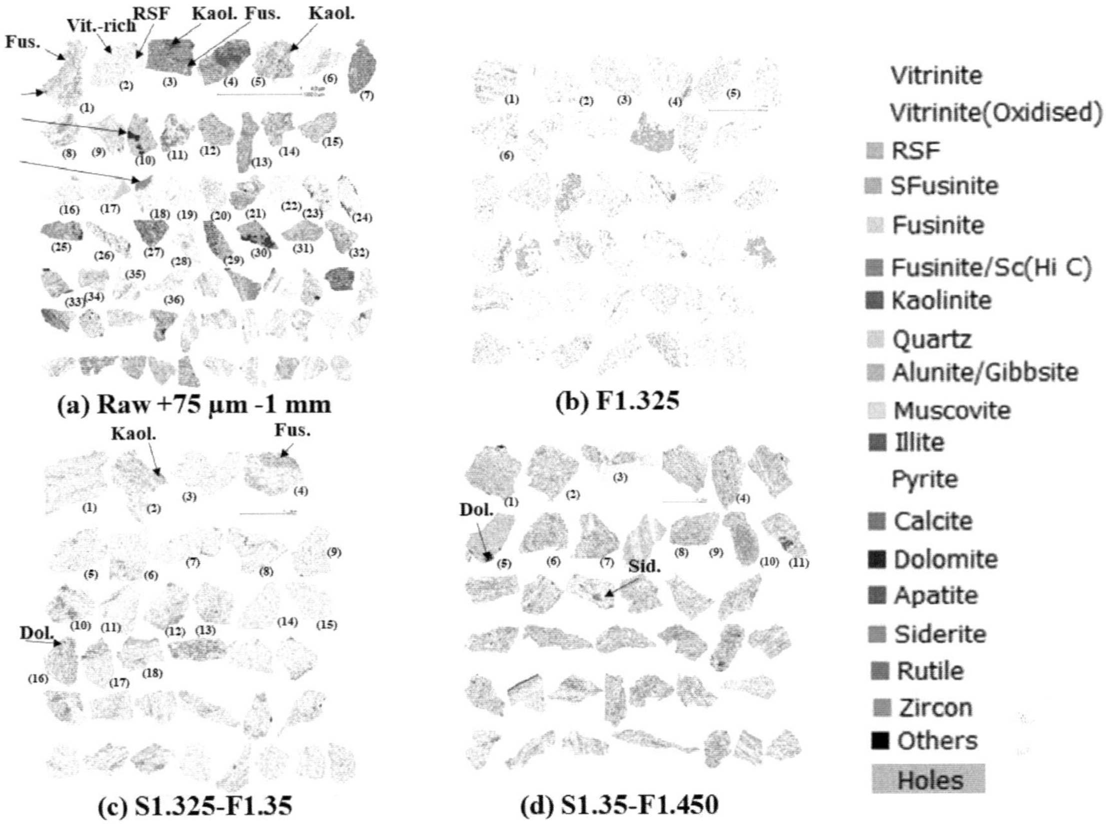

Fig. 2. SEM particle map results for Coal A raw and derived Float-sink sample fractions

Microlithotype particle classification for the raw coal and its float-sink fractions is in **Fig. 3**. The raw coal contained diverse microlithotypes: 26.32% (by mass) Vitrite80, 21.52% Vitrite60, 14.95% VITSF, 13.59% FUS, and negligible others. Tight density range float-sink processing resulted in significant enrichment of Vitrite80 in F1.325 (96.37%). The concentration of Vitrite80 particles declined with higher density fractions. F1.45 fraction had 53.32% Vitrite60, an improvement.

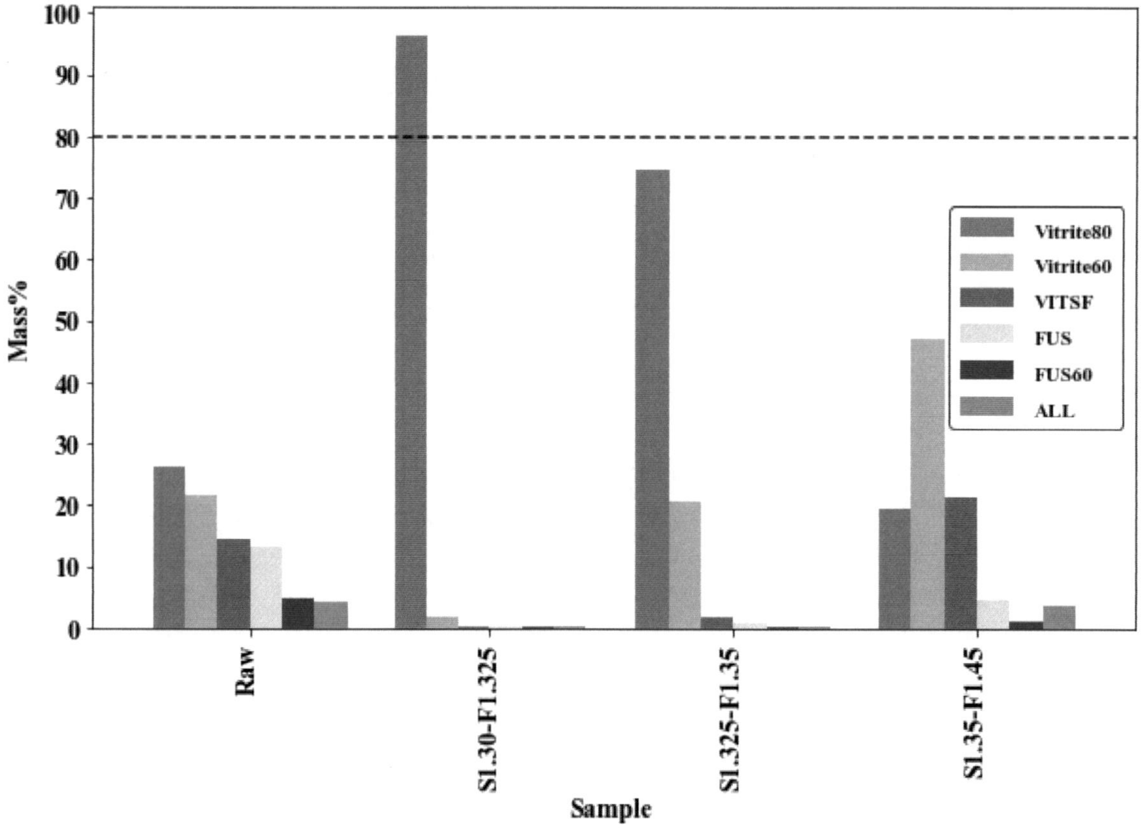

Fig. 3. SEM microlithotype particle type classification distribution results (maceralic group) for Coal A raw and derived Float-sink sample fractions

SEM particle maps for Coal B (Polish) raw and the derived Reflux Classifier fractions are shown in **Fig. 4 (a) to (b)**. SEM particle maps for Coal B raw contained fewer mineral-rich particles than Coal A (South African export power station feed). Coal B Reflux Classifier (RC) 1500 sample showed more homogeneous vitrinite-rich particles compared to the raw. Microlithotype particle classification results for Coal B raw and Reflux Classifier fraction RC 1500 are shown in **Fig. 5**. Coal D raw had high Vitrite80 particles (60.78%), which increased in the RC 1500 fraction to 80.51% due to its density facilitating separation from Mudstone particles.

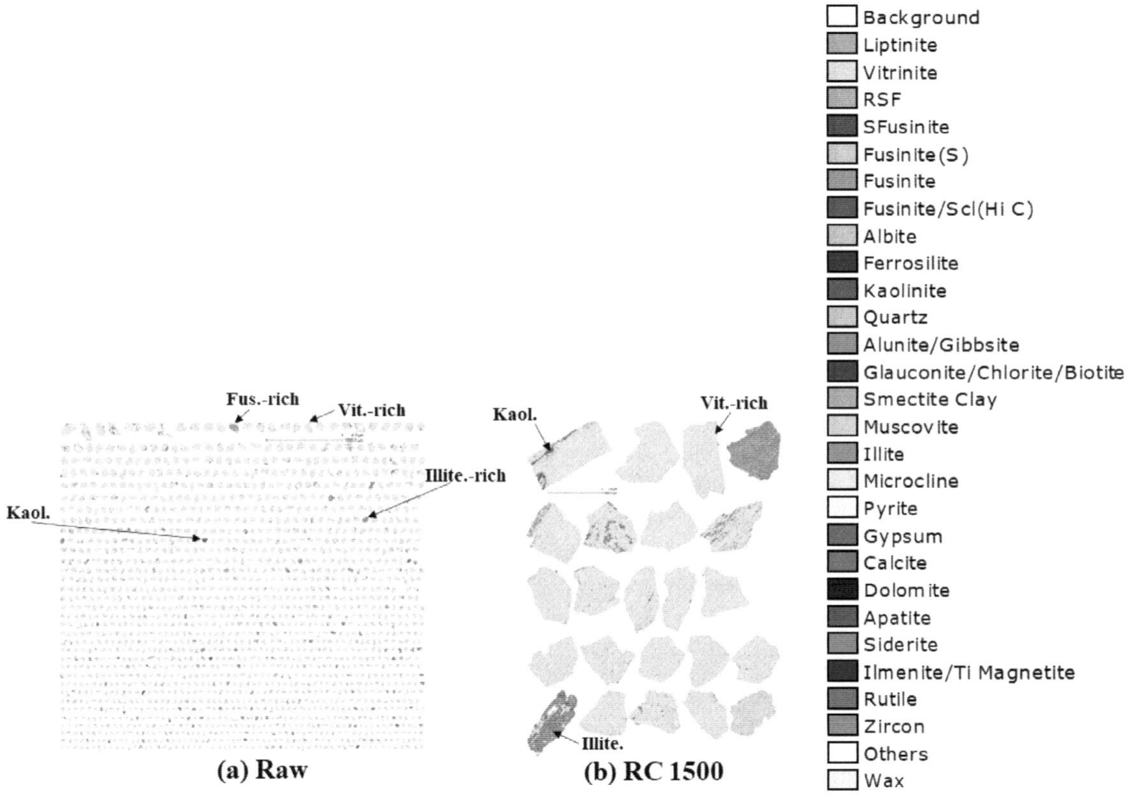

Fig. 4. SEM particle map results for Coal B raw and derived Reflux Classifier sample fractions

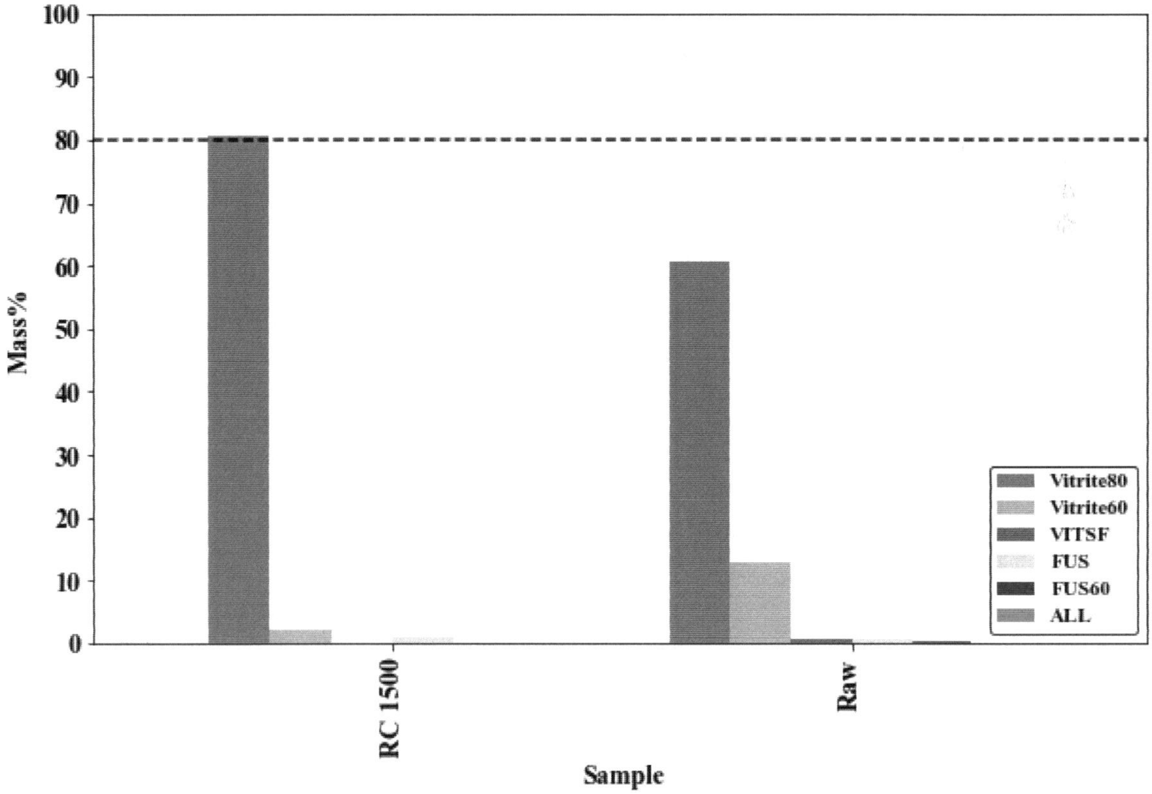

Fig. 5. SEM microlithotype particle type classification distribution results (maceralic group) for Coal B raw and derived Reflux Classifier sample fractions

3.2. DTF combustion tests and CFD modelling for chars derived from microlithotype particle type concentrates results

The predicted and experimental conversion results are shown in **Fig. 6 (a) to (f)**. The comparison indicates a good agreement between experimental and simulated results. The slight discrepancy between some measurement points and the model could be attributed to the solid residue collection efficiency and/or inaccuracies in the Ash Tracer Method. Coal A char combustion model results are compared in **Fig. 7** and results indicated that Coal A F1.325 (Vitrite80-rich) char particles ignited quicker and achieved 98% burnout quicker than the other Coal A samples. This is due to Coal A F1.325 char having a lower activation energy and ash yield than the other samples. The Coal A raw char was the least reactive compared to the F1.325, F1.350 and F1.450 char samples. A similar comparison was made for Coal B raw and RC 1500 char samples in **Fig. 7(e) to (f)**, indicating similar overlapping combustion profiles. The predicted model residence times were similar for the raw and fractions for Coal A and Coal B char samples. The model particle temperature was similar for both raw and fractionated chars of Coal A and Coal B. The kinetic parameters used in the Surface Reaction Model are provided in **Table 1**. Coal A F1.325 has the lowest activation energy compared to the other samples due to the sample having the highest proportion of Vitrite80 particles. The model includes the effect of water cooled injection probe which cools the secondary air (**Fig. 8 (a)**). This is likely to impact the ignition of char particles exiting the injection probe. Additionally, gas velocity, char reaction rate, heat reaction, and gas species for carbon dioxide and oxygen are detailed in **Fig. 8**. The mole fraction of oxygen is 3.0% (volume) at the reactor inlet but decreases as combustion proceeds and oxygen is consumed.

CFD model was simulated with two particle injections comprising of 75% (by mass) Vitrite80 and 25% (Vitrite60) and compared to Coal A F1.350 measured data in **Fig. 7 (c)**. A mass weighted profile was determined based on the proportion of Vitrite80 and Vitrite60 particles. The comparison indicates the mass weighted average profile does trend the measured results. This suggests the average performance of the whole coal sample could be predicted from the blend of its constituent coal microlithotype particle types. Further enhancement to the CFD model could include complex char combustion reactions. This CFD model could be used to assess combustion burnout characteristics of new coal sources and blends using its proximate, ultimate and SEM results. The same approach could potentially be also applied to other combustible particles such as biomass blends with coal.

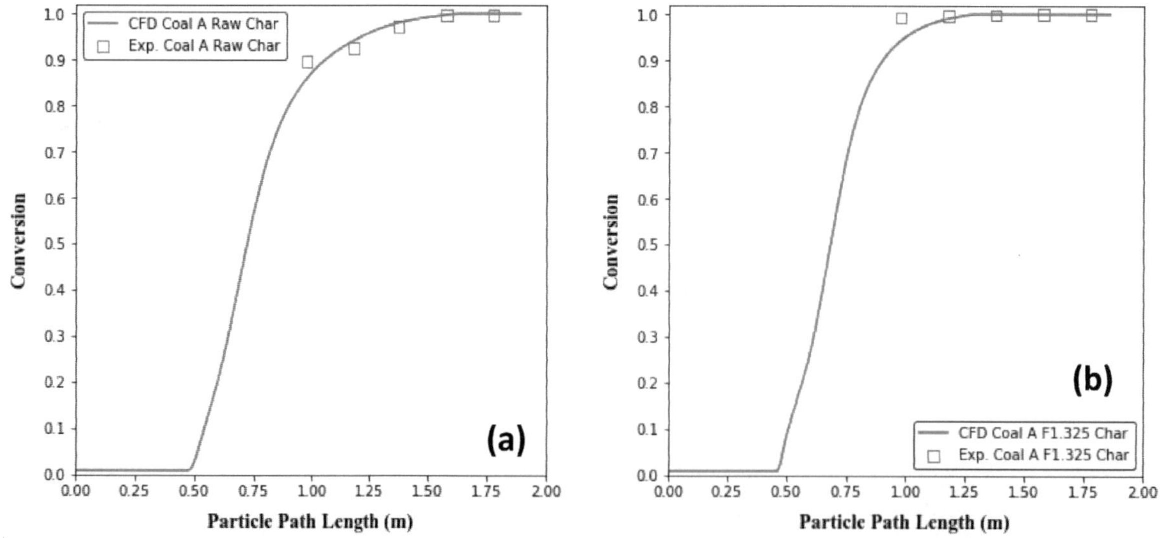

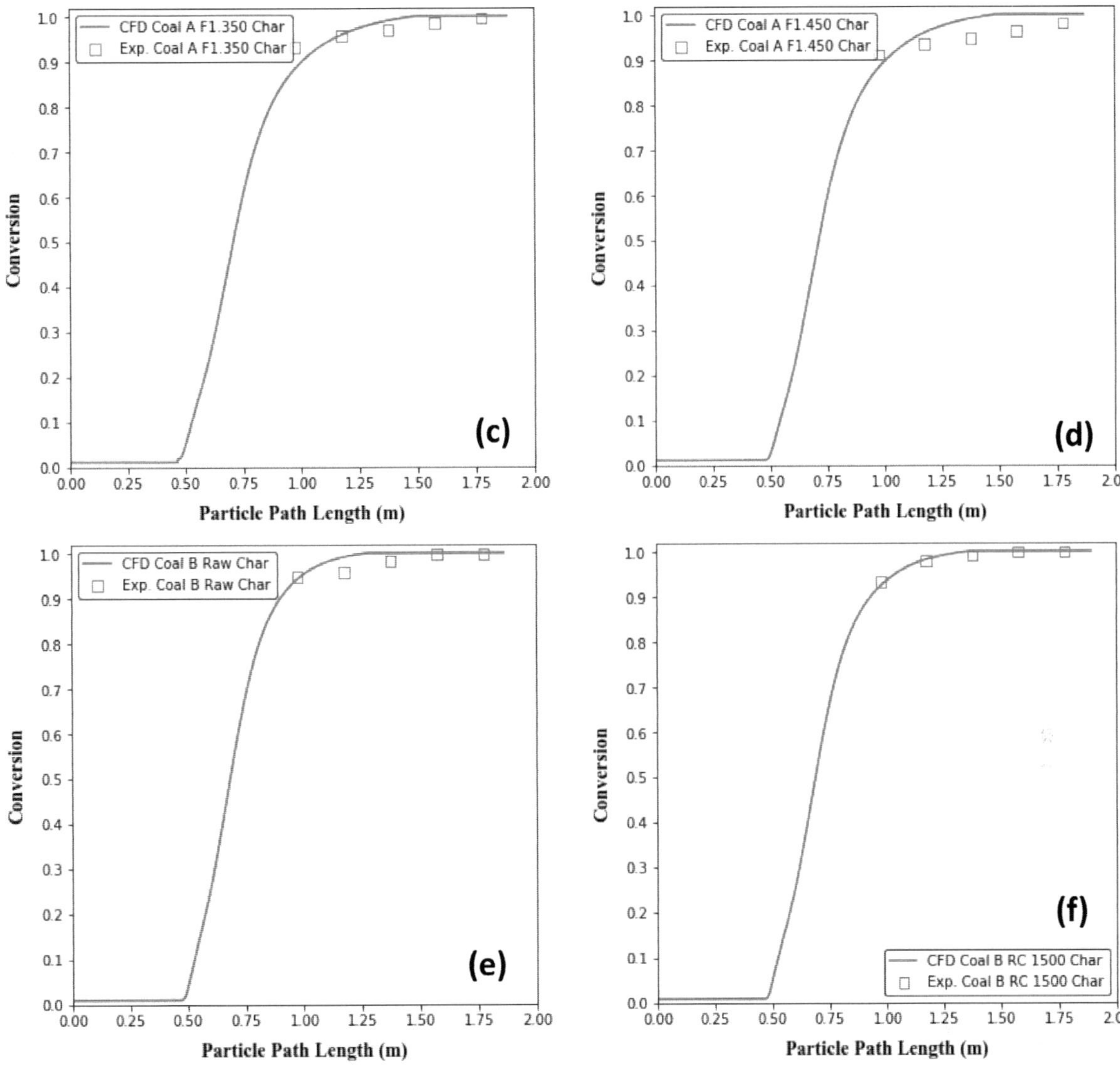

Fig. 6. DTF coal char combustion tests conversion and CFD simulation results, (a) CFD model results for Coal A Raw (SA Export), (b) F1.325, (c) F1.350, (d) F1.450, (e) Coal B Raw (Polish Export) and (f) Coal B RC 1500

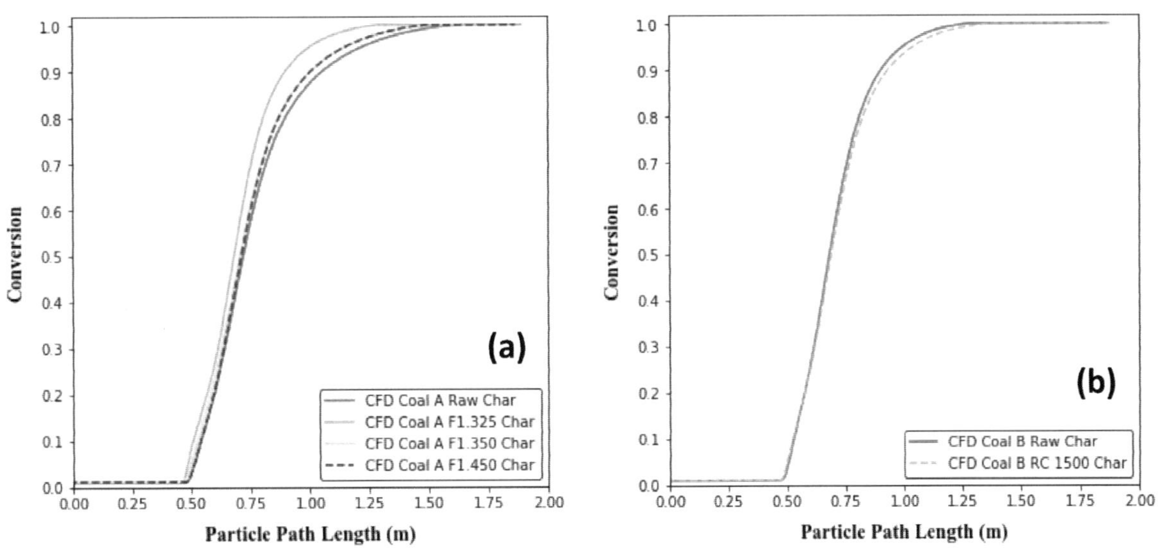

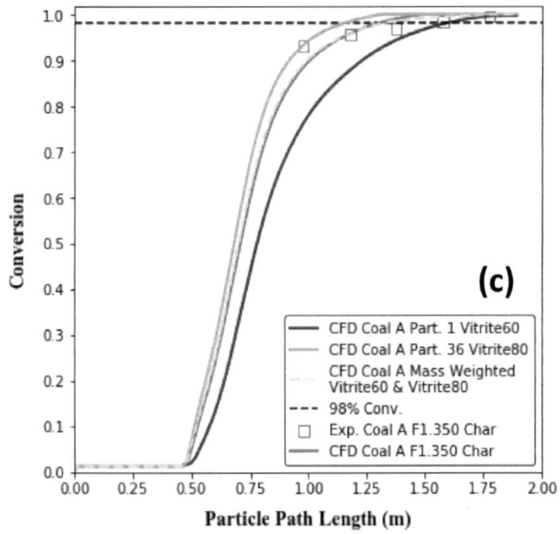

Fig. 7. DTF CFD coal char combustion simulation results, (a) CFD model results Coal A Raw (SA Export), F1.325, F1.350, and F1.450, (b) Coal B Raw (Polish Export) and Coal B RC 1500, (c) Coal A 75% Vitrite80, 25% Vitrite60, Mass Weighted Vitrite80 & Vitrite60, Coal A F1.350 experimental & simulation

Table 2: Summary of Surface Reaction Model kinetic parameters

No.	Sample Name	Activation Energy (MJ/kmol)	Pre-exponential Factor (1/s)
1	Coal A Raw	35	0.002
2	Coal A F1.325	20	0.002
3	Coal A 1.350	30	0.002
4	Coal A F1.450	41	0.002
5	Coal B Raw	35	0.002
6	Coal B RC 1500	30	0.002

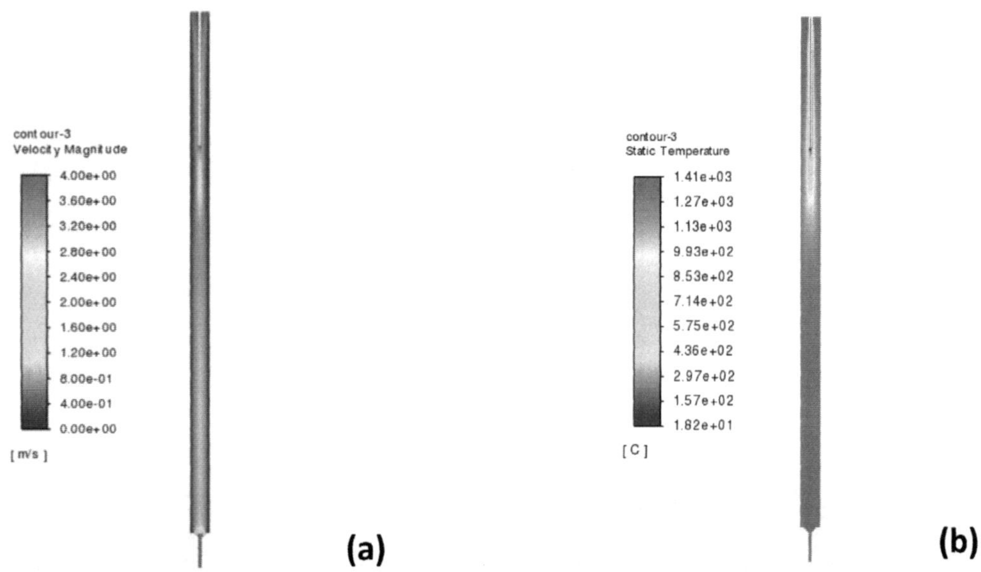

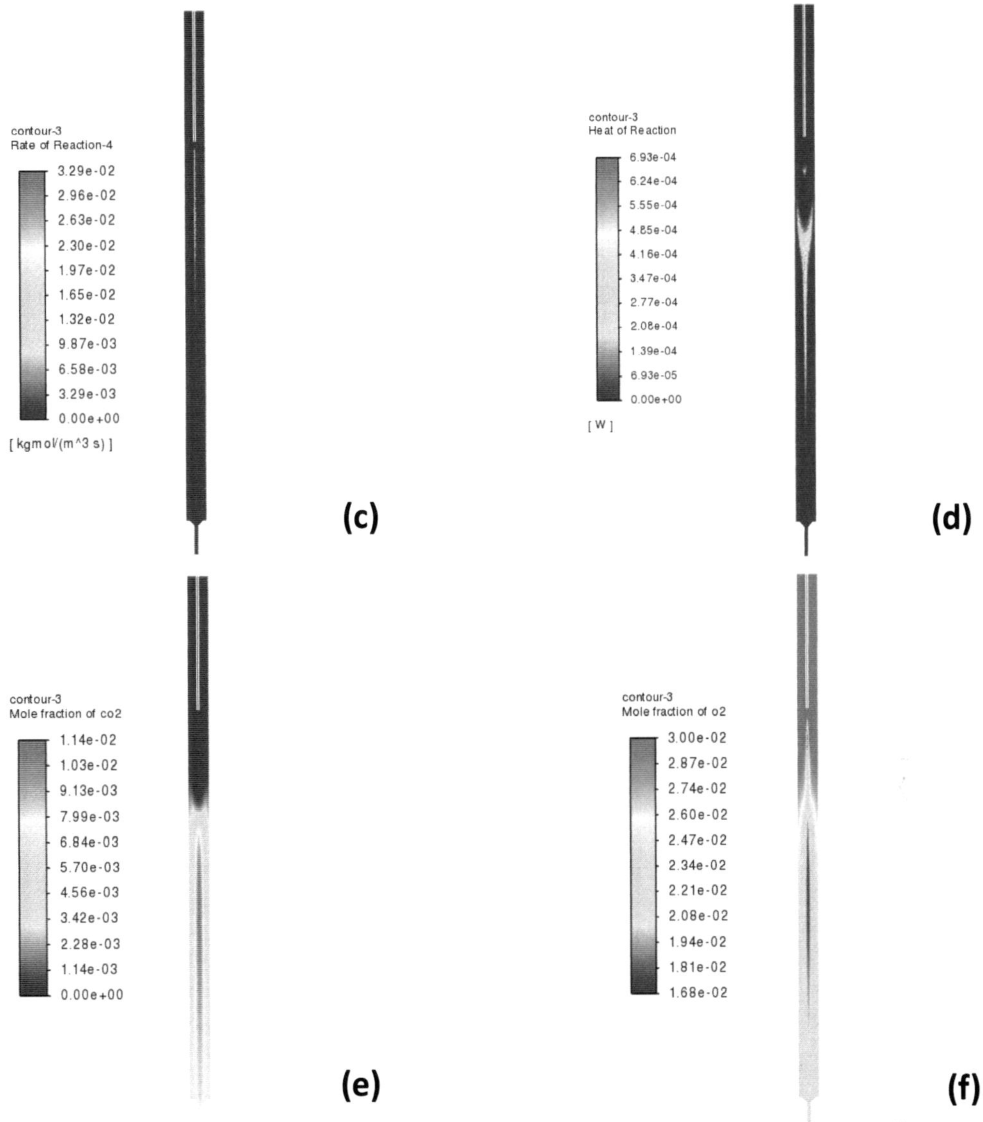

Fig. 8. DTF CFD coal char combustion simulation results, (a) gas velocity, (b) gas temperature, (c) char rate, (d) heat reaction, (e) gas carbon dioxide, (f) gas oxygen

4. Conclusions

The conclusions from this study on the DTF char combustion characteristics of coal microlithotype particle type concentrates and the CFD combustion modelling are as follows:

- The coal chars derived from Vitrite80 microlithotype particle types are more reactive, ignite quicker and have a shorter burnout residence time compared to the other microlithotype samples.
- The average burnout characteristics of the whole coal sample can be predicted accurately using CFD modelling and simulating the individual microlithotype particle types that constitute the sample.

5. Recommendations

The recommendations of this study are as follows:

- DTF pyrolysis tests should be conducted for coal microlithotype particle type concentrates to determine the reaction kinetic parameters and included in the CFD model.
- Include a detailed char reaction mechanism in the Surface Reaction Model.

- The developed CFD combustion model should be extended to other fuel particles, such as biomass co-firing with coal.

Acknowledgements

The authors acknowledge the financial support from Eskom Research, Testing, and Development and the support provided by the Eskom Power Plant Engineering Institute (EPPEI). They extend their gratitude to individuals who contributed significantly to this study: Bonny Nyangwa, Dr Chris Van Alphen and Priven Rajoo for technical guidance, Shinelka Singh and Thulisile Sishuba for support with QEMSCAN and TESCAN analyses, Prof Rosemary Falcon, Dr Samson Bada, and Dr Mark Van der Riet for advice during PhD meetings, Patrick Musie, Phumzile Mabidi, Xolani Ngubeni, and Olifant Sinthumule for chemical analysis aid, Mandla and John from Siza Laboratories for sample preparation, Sean Harbinson and colleagues at Bureau Veritas Laboratories for float-sink processing advice, Boitshoko Mashike, Bongani Khumalo, Molly Minyuku, Rendani Tshotheli and Prof Schmitz for export coal qualities assistance, and Jeffrey Mesolo for sample collection aid, Bonny Nyangwa and Kganuwi Kekana for their assistance with the DTF tests and providing advice. Reshendren Naidoo, Dr Frikkie Conradie, Prof Walter Schmitz, Lesego Moumakoe, Dr Falah Alobaid, and Dr Hrusikesh Barik for their assistance with Ansys Fluent CFD modelling. Nigel Dhamini for setting up and providing access to CFD workstations.

References

[1] X. Prevost, "Current and future use of coal in South Africa," in *Fossil Fuel Foundation Energy Conference*, 2015.

[2] DMRE, *Department of Mineral Resources and Energy Intergrated Resource Plan IRP2019,* 2019.

[3] DMRE, *Integrated Resource Plan 2023,* 2024.

[4] IEA, "Coal 2022 Analysis and forecast to 2022," 2022.

[5] I. Suarez-Ruiz and C. R. Ward, Applied coal petrology, 2008.

[6] M. Jakob, J. C. Steckel, F. Jotzo, B. K. Sovacool, L. Cornelsen, R. Chandra, O. Edenhofer, C. Holden, A. Lochel, T. Nace, N. Robins, J. Suedekum and J. Urpelainen, "The future of coal in a carbon-constrained climate," 2020.

[7] S. Ivanova, A. Vesinina, N. Fotina and A. Prosekov, "An overview of carbon footprint of coal mining to curtail greenhouse gas emissions," 2022.

[8] L. Moodley and P. Rajoo, "Duvha Power Station Unit 1 Corrosion Investigation," 2012.

[9] M. Blenkinsop, G. Twala, P. Gildenhuys and W. Van Remoortere, "Matimba Lower Seams Phase 3," 1994.

[10] D. Fong and H.-C. Magel, "Matimba Furnace Simulation," 1997.

[11] P. Rajoo, "Matla Seam 2 and 4 coal tests at the Pilot Scale Combution Test Facility," 2002.

[12] L. Bartonova, "Unburned carbon from coal combustion ash: An overview," 2015.

[13] M. Torresi, B. Fortunato, S. M. Camporeale and A. Saponaro, "CFD Modeling of pulverised coal combustion in an industrial burner," 2012.

[14] J. Pallares, I. Arauzo and A. Williams, "integration of CFD codes and advanced combustion models for quantitative burnout determination," 2007.

[15] R. Vuthaluru and H. B. Vuthaluru, "Modelling of a wall fired furnace for differenct operating conditions using Fluent," 2006.

[16] M. Xu, J. Azevedo and M. G. Carvalho, "Modelling of front wall fired utility boiler for different operating conditiosn," 2001.

[17] Z. F. Tian, P. J. Witt, M. P. Schwarz and W. Yang, "Numerical modelling of pulverised coal combustion," in *Handbook of multiphase flow science and technology*, 2023, p. 339.

[18] M. B. Gandi, R. Vuthaluru, H. Vuthaluru, D. French and K. Shah, "CFD based predition of erosion rate in large scale wall-fired boiler," 2012.

[19] C. Sheng, B. Moghtaderi, R. Gupta and T. F. Wall, "A computational fluid dynamics based study of the combustion characteristics of coal blends in pulverised coal-fired furnace," 2004.

[20] A. Williams, R. Backreedy, R. Habib, J. Jones and M. Pourkashanian, "Modelling coal combustion: The current position," 2002.

[21] L. Moodley, "Lethabo Power Station Units 2 & 3 combustion tests," 2013.

[22] J. C. Crelling, E. J. Hippo and B. A. Woerner, "Combustion characteristics of selected whole coal and macerals," 1992.

[23] N. Soundarrajan, N. Krishnamurthy and S. V. Pisupati, "Physical and Chemical Characterization of Coal Particles Used as Entrained Flow Gasifier Feedstock: Heterogeneity in Mineral Matter Distribution," 2012.

[24] G. Marangwanda, D. Madyira and T. Babarinde, "Coal Combustion models: A review," 2019.

[25] R. Falcon and A. Ham, "The characteristics of South African coals," 1988.

[26] P. Rajoo, "The application of pilot scale coal evaluation to full scale boilers," 2010.

[27] S. Singh, "The potential of QEMSCAN in predicting washability of Mpumalanga coals," 2015.

[28] J. Yu, J. A. Lucas and T. F. Wall, "Formation of the struture of chars during devolatilisation of pulverised coal and its thermoproperties A review," 2007.

[29] ICCP, "The new inertinite classification (ICCP System 1994)," 2001.

[30] R. H. Hurt and R. E. Mitchell, "On the combustion kinetics of heterogeneous char particle populations," 1992.

[31] J. Gibbins, T. Beeley, J. Crelling, A. Scott, N. Skorupska and J. Williamson, "Observations of heterogeneity large pulverised coal particles," 1999.

[32] C. van Alphen, Interviewee, *QEMSCAN coal particle type characterisation.* [Interview]. 4 April 2019.

[33] G. H. Taylor, M. Teichmuller, A. Davis, C. F. K. Diessel, R. Littke and P. Robert, Organic petrology, 1998.

[34] C. van Alphen, *TESCAN coal microlithotype analysis setup,* 2021.

[35] R. Backreedy, L. Fletcher, L. Ma, M. Pourkashanian and A. Williams, "Modelling pulverised coal combustion using detailed coal combustion model," 2006.

[36] K. P. Galvin, "Options for washability analysis of coal - A literature review," 2006.

[37] K. Bergh, "Trace element concentration reduction by beneficiation of Witbank Coalfield," 2010.

[38] L. Moodley, F. Conradie, W. Schmitz, J. Mathews, H. Neomagus, C. Van Alphen and B. Nyangwa, "Microlithotype particle type classification and enrichment of coal maceral concentrates with Relfux Classfier, Float-Sink and RhoVol," 2024.

[39] S. Harbinson, *Float-Sink Test Methodology,* 2020.

[40] L. Moodley, "Development of a prognostic pulverised coal combustion procedure phase 1," 2022.

[41] Y. Lui, "Mineral matter organic matter association characterisation by QEMSCAN and application in coal utilisation," 2005.

[42] French, David; Ward, Colin R, "The application of advanced mineralogical techniques to coal combustion product characterisation," 2009.

[43] Ansys Fluent, "Ansys Fluent Theory Guide Version 16.2," 2015.

[44] R. M. S. Falcon and C. P. Synman, "An introduction to coal petrography: Atlas of Petrographic Consituents in the Bituminous coal of Southern Africa," 1986.